Emergency Vehicle Safety Initiative

FA-272 / August 2004

FEMA

U.S. FIRE ADMINISTRATION MISSION STATEMENT

As an entity of the Federal Emergency Management Agency, the mission of the United States Fire Administration is to reduce life and economic losses due to fire and related emergencies, through leadership, advocacy, coordination, and support. We serve the Nation independently, in coordination with other Federal agencies, and in partnership with fire protection and emergency service communities. With a commitment to excellence, we provide public education, training, technology, and data initiatives.

On March 1, 2003, FEMA became part of the U.S. Department of Homeland Security. FEMA's continuing mission within the new department is to lead the effort to prepare the Nation for all hazards and to manage Federal response and recovery efforts effectively following any national incident. FEMA also initiates proactive mitigation activities, trains first responders, and manages Citizen Corps, the National Flood Insurance Program, and the U.S. Fire Administration.

TABLE OF CONTENTS

PREFACE

The United States Fire Administration would like to acknowledge the following individuals and organizations who participated in one or more of the National Forums on Emergency Vehicle Safety or conducted the information gathering site visits. The participants formed a group with one of the most comprehensive bases of expertise in emergency vehicle response and highway scene safety ever assembled. The extensive information provided within this report would not have been possible without their dedication and efforts on this project.

Organization	Representative(s)
American Ambulance Association	Maria Bianchi
	Larry Wiersch
Congressional Fire Service Institute	Robert J. Barraclough
Cumberland Valley Volunteer Firemens Association	Steve Austin
	Harry Carter
Federal Aviation Administration Technical Center	Keith Bagot
Fire Apparatus Manufacturers Association	Glen McCallister
Fire Department of New York/Firehouse Magazine	Michael Wilbur
Fire Department Safety Officers Association	Charles Soros
General Services Administration	Mel Globerman
International Association of Fire Chiefs	Richard Knopf
International Association of Fire Fighters	Andy Levinson
	Richard Duffy
International Fire Service Training Association	Michael Wieder
IOCAD Emergency Services Group	Richard Gurba
	Neil Honeycutt
	Warren James
	Judy Janing (Principal Writer)
	Murrey Loflin
	Richard Marinucci
	Eric Nagle
	Kevin Roche
	Gordon Sachs
	Steve Weissman
	Michael Wieder (Project Manager)
Medical Transportation Insurance Professionals	William Leonard
Mitretek Systems (supporting FHWA USDOT-ITS)	Ken Brooke
National Association of Emergency Vehicle Technicians	Don Henry
	Jeff Dickey
National Fire Protection Association	Gary Tokle
	Carl Peterson
National Institute for Occupational Safety & Health	Nancy Romano
National Institute of Standards & Technology	Nelson Bryner
National Safety Council	Barbara Caracci
National Truck Equipment Association	Mike Kastner
National Volunteer Fire Council	Bob Cumberland
North American Fire Training Directors	Butch Weedon
Plano, Texas, Fire Department	Alan Storck
Training Resources and Data Exchange	Jim White
	Art Cota
U.S. Army TACOM	David Tenenbaum
U.S. Department of Transportation/National Highway Traffic Safety Administration	David Bryson
	Craig Allred

U.S. Fire Administration   Cathy Broughton
                                                               Charlie Dickinson
                                                               Jeff Dyar
                                                               Alex Furr
                                                               Kirby Kiefer
                                                               Bill Troup
Volunteer Fireman's Insurance Services (VFIS)   David Bradley
                                                               William Jenaway
                                                               Mike Young
                                                               Richard Patrick

The following people are acknowledged for their photographic contributions to this report:
    Leland Bishop
    Judy Janing
    Murrey Loflin
    Eric Nagle
    Alan Storck
    Steve Weissman
    Michael Wieder

The United States Fire Administration would also like to acknowledge the tremendous support and assistance provided by the U.S. Department of Transportation (DOT) Intelligent Transportation Systems (ITS) Joint Program Office.

INITIATIVE OVERVIEW

INTRODUCTION

As traffic volume increases and the highway and interstate system becomes more complex, emergency responders face a growing risk to their personal safety while managing and working at highway incidents. The purpose of this report is to identify practices that have the potential to decrease that risk, as well as to reduce the number of injuries and deaths that occur while responding to and returning from incidents (Figure 1-1).

Since 1984, 20 to 25 percent of firefighter fatalities annually resulted from motor vehicle crashes. From 1990 to 2000, 18 percent of the fatalities occurred responding to an alarm, and 4.1 percent occurred returning from an alarm. Of the firefighters who died in motor vehicle crashes, 25 percent were killed in privately owned vehicles (POV's.) Following POV's, water tankers, engines, and airplanes were most often involved in fatal crashes. Tankers claimed more fatalities than engines and aerial apparatus combined. Approximately 27 percent of fatalities were ejected at the time of the crash. Only 21 percent were reportedly wearing restraints prior to the crash. Table 1-1 (on page 2) provides a summary analysis of fatal emergency vehicle crashes that occurred while responding to and returning from an alarm from 1994 to 2001.

Figure 1-1. Springfield, Virginia, interchange.

According to a study by Maguire, Hunting, Smith, and Levick, emergency medical services (EMS) personnel in the United States have an estimated fatality rate of 12.7 per 100,000 workers.[1] There is no comprehensive central database for ambulance crash data, but some estimate 15,000 crashes per year. A *USA Today* analysis of the National Highway Traffic Safety Administration (NHTSA) fatal, multivehicle ambulance crash data between 1980 and 2000 revealed the number killed in the other vehicle was 21 times greater than the number of ambulance drivers who died.[2]

According to the NHTSA Fatality Analysis Reporting System (FARS), there were 300 fatal crashes involving ambulances from 1991 to 2000. These 300 crashes resulted in 82 deaths of ambulance occupants and 275 deaths of occupants of other vehicles and pedestrians. FARS does not differentiate ambulance workers from passengers among those experiencing nonfatal injuries in fatal crashes. However, seating positions for all occupants and the severity of injuries can be determined. Using the FARS "fatal injury at work" variable with death certificate information, calculations indicated that 27 occupant fatalities were on-duty EMS workers.[3]

According to the National Emergency Medical Services Memorial Foundation in Roanoke, Virginia, 3 of the 16 EMS worker fatalities (19 percent) in 2001 to 2002, resulted from vehicle-related causes; one was involved in a crash and two were struck by vehicles while at the scene of an accident.[4]

An 11-year retrospective analysis of the characteristics of fatal ambulance crashes by Kahn, Pirrallo, and Kuhn revealed that the majority of ambulance crashes occurred during the emergency mode of operation and rear compartment occupants were more likely to be injured than those in the front compartment.[5] The findings of

TABLE 1-1.
FIREFIGHTER FATALITIES RESPONDING TO AND FROM AN ALARM: 1994-2001

	1994	1995	1996	1997	1998	1999	2000	2001
Total Fatalities	12	14	20	15	14	14	20	23
Apparatus Collision	(12)	(14)	(14)	(13)	(12)	(11)	(18)	(20)
Falls	(0)	(1)	(2)	(3)	(2)	(2)		
Vehicle Type								
Fire Apparatus:								
Engine	2	5	4	2	4	5	3	3
Water Tanker	4	2	1	2	1	4	3	5
Other (including ambulance)	3		2	3	3	2	8	6
Private Vehicle	1	5	7	5	4	2	6	9
Unknown	1		5	2	1			
Contributing Factors*								
Behavior – Fire Personnel	1	4	4	4	3	6	9	9
Behavior – Civilians	1		3	2	2	1	3	4
Technical Problems	1	1			1	1	1	3
Laws/Policy	1	4		1	5	4	9	8
Other (Environment, weather, medical condition)	2	1	5	3		1	6	12
Unknown	7	3	7	6	3	1	1	

* Fatality may have more than one contributing factor noted, thus, numbers will not sum to total number of fatalities.
Source: U.S. Fire Administration, *Fire Fighter Fatality Reports* (1994-2001).

the NHTSA analysis appear to contradict the findings of Kahn, Pirrallo, and Kuhn. The NHTSA analysis revealed that 11 (41 percent) of worker fatalities occurred in the driver's seat and 5 (19 percent) in the front passenger seat. Although not coded specifically, seven (26 percent) probably occurred in the patient compartment and four were unknown. The NHTSA analysis did reveal a greater injury severity associated with riding in the patient compartment, which supports the findings of Kahn, Pirrallo, and Kuhn.[3]

As with firefighter fatality data, non-use of restraints is an important contributing factor to EMS worker fatalities. Seven (26 percent) were drivers not wearing a restraint, two were unrestrained in the front passenger seat, and six were unrestrained in the patient compartment.[3] This is due in part to the design of lap belt restraint systems commonly found in the patient compartment. These systems do not allow full access to the patient. If the belt is used properly the responder is positioned against the side wall, making it impossible to bend forward to access the patient or the cabinets along the wall on the driver's side. The belts must be unbuckled for the responder to stand over or kneel near the cot to perform certain treatments.[6] This limitation results in many responders sitting on the edge of the side bench and riding unrestrained.

BACKGROUND

The consistently high annual percentage of emergency worker fatalities related to response prompted the *Fire Service Emergency Vehicle Safety Initiative* (EVSI), a partnership effort among the U.S. Fire Administration (USFA), the U.S. Department of Transportation (DOT)/NHTSA, and the DOT/Intelligent Transportation Systems (ITS) Joint Program Office. One of the primary functions of the EVSI was to sponsor the National Forum on Emergency Vehicle Safety, which brought together representatives of major national-level fire and emergency service associations and other individuals and organizations with an interest and expertise in emergency vehicle safety. These representatives met three times between June 2002 and August 2003. The purpose of these meetings was to:

- identify the major issues related to firefighter fatalities that occur while responding to or returning from alarms and while operating on highway emergency scenes;
- develop and prioritize recommendations to reduce firefighter response and highway scene fatalities;
- identify organizations that had made progress in improving firefighter/responder safety in these areas based on mitigation techniques and technologies; and
- review and approve the findings of the research done for this report.

METHODOLOGY

During the first meeting, the participants formed several focus groups that identified multiple issues related to emergency response and highway scene safety. Potential recommendations to address each issue were also listed. Following the meeting, these issues and recommendations were categorized.

At the second meeting, participant focus groups reviewed, refined, and expanded the recommendations from the first meeting in relation to technology, training, and operational aspects as follows.

Legislation, Standards, and SOP's

- Develop model legislation based on information from the National Committee on Uniform Traffic Laws and Ordinances and the Model Uniform Traffic Code (MUTC) for both fire apparatus and civilian drivers.
- Review the MUTC for pertinent information to help develop model Standard Operating Procedures (SOP's) and legislation.
- Develop model SOP's and enforcement criteria.
- Develop driver qualifications. (These should be verified before allowing a person to drive.)
- Implement available standards and requirements for marking highway emergency scenes, such as those of the American National Standards Institute (ANSI), the International Safety Equipment Association (ISEA), and the DOT.
- Require the use of compliant protective vests when operating on the highway.
- Penalize violators of highway safety practices/requirements.
- Submit public comment for NFPA 1901, *Standard on Automotive Fire Apparatus*, encouraging the use of brightly colored seatbelts, marking vehicle dimensions, and the use of reflective markings for vehicle compartments and interior aspect of doors.

Since public comments were closing soon after the second meeting, the forum participants voted to submit these comments as participants of the EVSI, not endorsed by USFA. There were five public comments submitted to the NFPA Fire Department Apparatus Committee. Appendix A contains the submitted wording, the final wording that went into 1901, and the committee's substantiation for its changes to the submitted wording.

Training

- Develop a layered (modular) training program that includes specific training for all levels of an organization on their responsibilities related to emergency vehicle safety. Modules would be developed for responders, drivers, company officers, chief officers, emergency vehicle technicians (mechanics), instructors, educators, and local government officials.
- Develop and implement legitimate proficiency testing to assess the performance of drivers.

- Develop a certification/recertification program as an extension of the layered training program and proficiency testing process. It could be based on the commercial driver license (CDL) process and should be similar to the red card program for wildland certification.
- Put training and testing in the context in which it actually will be used.

Technology
- Assess opportunities to expand the use of traffic control devices.
- Assess the effectiveness of:
  - Orange or yellow seatbelts,
  - Seatbelt extensions and brightly colored reception ends,
  - Seatbelt lights at the officer's position and above an unbelted, occupied seat,
  - Passive restraints (airbags),
  - Remote control mirrors,
  - Lateral acceleration warning devices, and
  - Vehicle visibility enhancements on apparatus and personnel protective equipment.

General Fire Service Education
- Implement an awareness campaign to educate emergency service personnel about the seriousness of the emergency vehicle safety issues, POV response issues, and the need for a change in behavior and attitude. This awareness must occur on all levels.
- Train leaders about the aspects they are responsible for; advocate leadership responsibility for safety.
- Increase safety awareness of drivers' supervisors.
- Increase awareness of safety issues related to unrestrained objects in the cab.
- Increase knowledge of vehicle modifications that affect vehicle stability.

Delivery of Safety Information
FIRE SERVICE
- Develop a Web site for organizations to share and download model SOP's, progressive programs, and other safety information.
- Develop regional safety program conferences.
- Review the National Fire Academy (NFA) curriculum to determine the amount/level of highway safety information in various curricula. At a minimum, it should be included in the Health & Safety and Public Education curricula.
- Develop informational pieces for the USFA Web site.

PUBLIC
- Develop an awareness campaign to educate the public on how to react properly when approaching/being approached by an emergency vehicle, and how to traverse safely through an emergency scene.

DATA COLLECTION
- Develop a national database on emergency vehicle crashes and any resulting injuries/deaths to both firefighters and civilians.
- This could begin by enhancing an existing database (adding a 'box' for involvement of emergency vehicle and type). The new system should be based on the National Fire Incident Reporting System (NFIRS) technology and programming so it can be integrated easily into the fire service. The development of the database must involve law enforcement, NHTSA, and USFA.
- Implement a near-miss reporting system, similar to the 'Green Sheet' program

by the California Department of Forestry and Fire Protection (CDFFP).

- Encourage research on highway safety-related issues in the Executive Fire Officer (EFO) program at the NFA.

ADDITIONAL RECOMMENDATIONS

- Incentives should be developed by the insurance industry and other sources to encourage those departments working toward improved emergency vehicle safety.

- Continue and expand the work of the participants in the National Forum on Emergency Vehicle Safety as the National Emergency Vehicle Safety Consortium under the EVSI project. The consortium should be modeled after the National Fire Service IMS Consortium and would be responsible for such things as:

  — Developing model SOP's,

  — Developing model State laws,

  — Identifying new technologies, and

  — Developing publications.

Forum participants recognized that it would be impossible to address all of these issues and areas in a reasonable amount of time. Therefore, they selected five topical areas to identify progressive programs that would be easy to implement and would help emergency response agencies make immediate improvements to their procedures. Based on personal knowledge and experience, participants also recommended agencies for site visits to examine progressive programs in these areas. These five basic topical areas and the agencies that were selected for review are listed in Table 1-2.

An interview team visited each department. Specific questions, developed prior to

Topic	Aspects Reviewed	Site
TABLE 1-2. PARTICIPANT-IDENTIFIED REVIEW SITES		
Emergency Vehicle Driver Training	• Effectiveness of a "stepped" driver training program • Benefit of apparatus operator certification program • Use of emergency vehicle training simulators	CDFFP, Ione, CA Ventura County, CA Sacramento Regional Training Facility, CA
Vehicle Marking On-Scene Warning Lights	• Reflective marking patterns • Warning light types/colors/patterns	Fairfax County, VA Phoenix, AZ Plano, TX
Highway Scene Markings	• Use of lights to warn other vehicles approaching an emergency scene	Fairfax County, VA Plano, TX Virginia Beach, VA
Priority Dispatch Response SOP's	• Types and patterns of lighting • Apparatus positioning • Placement of stand-alone barriers, markings, and lane-closure devices • Policies that limit emergency mode (lights and sirens) response based on the type and severity of the reported emergency	Virginia Beach, VA Salt Lake City, UT Phoenix, AZ Salt Lake City, UT Virginia Beach, VA
Optical Pre emption	• Effects of optical preemption devices on response times, vehicle crashes/near misses, and integration in the intelligent transportation system	

the visit, served as the basis and starting point for each interview. Department personnel often provided additional information related to other safety practices. After the site visit, the interview notes were reviewed, collated, and summarized by topic. All team members reviewed the summary for accuracy. This information was used to compile the report draft. Forum participants reviewed this draft and added final recommendations in August 2003.

DEFINITION

There are several terms used to identify personnel who drive apparatus, including engineer, operator, and driver. In the context of this report, the term "driver" is used to identify personnel who have the responsibility of actually driving the vehicle.

LIMITATIONS

It must be stressed that there are three major limiting factors to the findings of this report:

1. There is no comprehensive database to identify factors involved in emergency vehicle crashes, emergency worker injuries, or emergency worker fatalities.

2. The databases that do exist may or may not include information related to either volunteer fire or EMS services.

3. Very few of the departments visited gather any data to evaluate the effectiveness of any of the practices reviewed.

Because of these limitations, the reader must keep in mind that the progressive programs discussed in this report are anecdotal. Although anecdotal in nature, they have been endorsed by the participants of the National Forum on Emergency Vehicle Safety. The reader must also remember that the success of the programs discussed in this report, as well as all safety programs, requires both a multifaceted approach and member and management commitment.

It is hoped that this initial report will stress the need to develop a comprehensive database and serve as the basis to identify areas for true scientific research that will identify a specific problem, define its scope, and collect specific data for analysis to determine the actual effects of a given practice or intervention on pre-identified outcomes. The outcomes for specific studies must be measurable over a relatively short timeframe, using a limited number of participating departments. Impact on fatalities cannot be measured in any one study, since the number of fatalities is small in a scientific sense and may not be accurate because of the limited database.

---

[1] U.S. Department of Health and Human Services, Centers for Disease Control and Prevention. (2003, February 28). Ambulance crash-related injuries among emergency medical services workers — United States, 1991-2002. *Morbidity Mortality Weekly Report*, 52 (08), 154-156.

[2] Davis, R. (2001, March 21). Speeding to the rescue can have deadly results. *USA Today*.

[3] U.S. Department of Health and Human Services, Centers for Disease Control and Prevention. op cit.

[4] Harvey, N. (2003, May 25). Service honors fallen EMS providers. *Roanoke Times*.

[5] Kahn, C.A., Pirrallo, R.G., Kuhn, E.M. (2001). Characteristics of fatal ambulance crashes in the United States: an 11-year retrospective analysis. *Prehospital Emergency Care*, 5, 261-269.

[6] U.S. Department of Health and Human Services, National Institute for Occupational Safety and Health. (2001). 26-year-old emergency medical technician dies in multiple fatality ambulance crash—Kentucky. Pub. No. FACE-2001-11.

AGENCY PROFILES

A total of 12 agencies were visited in the process of gathering information on the progressive programs selected for review. Following are the profiles of each of these agencies.

CALIFORNIA DEPARTMENT OF FORESTRY AND FIRE PROTECTION

The CDFFP employs 2,500 full time personnel. During fire season, the department adds an additional 2,500 seasonal workers, bringing the total to approximately 5,000. The department currently has nine different first-line engine models. There are 336 first-line engines, 36 to 40 reserve engines, and 40 camp engines. Last year the department used an additional 10 augmentation engines. These are engines that have been replaced, but remain in service until fire season is over. The use of augmentation engines is based on need and is not a standard practice. There are also 19 engines at the Training Academy. These have been used in the past to form three strike teams. The department is in the process of moving to 100 percent fully enclosed crew cabs to replace the engines that still require the crew to ride exposed in the back. At the time of this writing, the cost of insurance for each piece of apparatus was approximately $175 annually. The Office of Risk Insurance Management (ORIMS) determines the cost based on the financial impact of crashes.

The CDFFP Academy was established in 1967 and is located in Ione, California. Of the 19 engines at the academy, 4 are structural Type I and the rest are wildland Type III. The staff trains approximately 300 potential fire apparatus drivers each year. Approximately 200 are trained at the academy and 100 are trained off-site. The academy has classroom facilities, a driving course, practice areas for wildland vehicle operations, and a skid pan. It also has dormitory space and its own cafeteria. Students attending the academy reside in the dormitories until their training is complete.

Acknowledgements

The following are acknowledged for their expertise and willingness to give of their time in support of this project:

Dave Ebert	Chief, State Director of Fire Training
Paul Sans	Battalion Chief, Training Academy
Rick Brown	Senior Equipment Manager
Benny Aguilar	Limited Term Engineer (Driver)
Darren Dow	Limited Term Engineer (Driver)

FAIRFAX COUNTY (VIRGINIA) FIRE AND RESCUE

The Fairfax County Fire and Rescue Department is a combination career and volunteer organization providing fire suppression, emergency medical, technical rescue, hazardous materials, water rescue, life safety education, fire prevention, and arson investigation services. EMS includes advanced life support (ALS) response by paramedic engines and medic transport units. The department also sponsors a FEMA Urban Search and Rescue (US&R) Task Force (Virginia Task Force 1).

The department serves a population of 964,712 and covers a 399-square-mile area with a complement of 1,166 uniformed personnel and 12 active volunteer fire departments with 361 operational members. There are 35 stations with the following distribution of apparatus:

35 Paramedic engines	18 Ambulances
22 Medic units	8 Rescues
12 Trucks	2 Foam units

In 2002, the department made 89,246 total responses, which are categorized as follows:

60,685 – EMS
23,579 – Suppression
 4,982 – Public service

Acknowledgements

The following are acknowledged for their expertise and willingness to give of their time in support of this project:

Charles Atkins	Captain, Safety Officer
Rescue 19	A Shift Personnel
Engine 24	A Shift Personnel
Station 8	A Shift Personnel
Carol Salinger	Captain, Safety Officer, Arlington County Fire Department

PHOENIX (ARIZONA) FIRE DEPARTMENT

The Phoenix Fire Department is one of the busiest fire departments in the United States. The department serves a population of 1,387,662 and covers a 485.9-square-mile area with a complement of 1,366 sworn personnel and 332 civilians. The department has 48 fire stations with the following distribution of apparatus:

41 ALS Engines	12 Brush trucks
13 BLS Engines	5 Tankers
13 Trucks	3 Utility
12 Ladder tenders	8 Special operations (3 Haz mat)
22 BLS Ambulances	7 Air & Foam
7 ALS Ambulances	10 Support vehicles

The department also sponsors a FEMA US&R Task Force (Arizona Task Force 1). In 2002, the department made 185,101 total calls, which are categorized as follows:

 16,339 – Fire
104,635 – EMS
    563 – Special operations
 11,177 – Miscellaneous
 52,387 – Transports

Acknowledgements

The following are acknowledged for their expertise and willingness to give of their time in support of this project.

Kevin Roche	Resource Management
Bill Bjerke	Apparatus Supervisor
Nick Brunacini	North Division Commander
Mike Worrell	Engine 18 Captain
Ryan Dooley	Engine 18 Driver
Kirk Hover	Engine 13 Captain
Nick Petrucci	Engine 13 Driver
Brian Parks	Safety Officer
Louise Smith	Dispatch Supervisor

PLANO (TEXAS) FIRE DEPARTMENT

The Plano Fire Department is a career organization. The department serves a population of 230,000 and covers approximately 70 square miles with a complement of 262 uniformed field and 32 support personnel. There are 10 stations with the following distribution of apparatus:

11 Paramedic Engines   6 Ambulances   4 Paramedic Trucks

In 2002, the department made 16,498 total responses, which are categorized as follows:

Fires		Other Emergency Calls	
Structures		Medical	9,035
Residential	148	(including motor vehicle crashes)	
Other	32		
Outside of structures	80	Hazardous conditions	724
Vehicular	101	Public service	897
Grass	77	Good intent	1,946
Dumpster/Refuse	76	False calls	2,559
Other	9	Undetermined	45
		Cancelled by CAD	1,005
TOTAL	523	TOTAL	16,211

The Plano Fire Department has established itself as a model for fire departments internationally. Plano Fire Department has received Accredited Agency status with the Commission on Fire Accreditation International (CFAI) for meeting the criteria established through the CFAI's voluntary self-assessment and accreditation program. The department is the second fire department in the country to receive accreditation from the Commission on Accreditation of Ambulance Services (CAAS) for its compliance with national standards of excellence.

The City of Plano is one of only 29 cities in the United States to hold a Class 1 Public Protection Classification from the Insurance Services Office, Inc. (ISO). The Plano Fire Department is the only ISO Class 1, CAFI-accredited and CAAS-accredited fire department in the United States.

Acknowledgements

The following are acknowledged for their expertise and willingness to give of their time in support of this project:

William Peterson	Fire Chief
Kirk Owen	Division Chief (Support Services and Operations)
Alan Storck	Section Chief (Training, Safety)
Tami Laake	Risk Management
Jay Benton	Captain, Station 2
Mark McKown	Fire Apparatus Operator (Driver), Station 2
Michael Covey	Firefighter, Station 2
Chris Mougia	Firefighter, Station 2
Billy Lay	Captain, Station 5
Allen Light	Lieutenant, Station 5
Jack Miller	Fire Apparatus Operator (Driver), Station 5
Eric Everson	Firefighter, Station 5
Robert Hogan	Firefighter, Station 5
Toby Peacock	Captain, Station 5
Les Ruble	Fire Apparatus Operator (Driver), Station 5
Jeffrey Deskeere	Firefighter, Station 5
Mark Doerr	Firefighter, Station 5
Shawn Childress	Firefighter, Station 5
Keil Baldia	Firefighter, Station 5

SACRAMENTO (CALIFORNIA) REGIONAL DRIVER TRAINING FACILITY

The Sacramento Regional Driver Training Facility was established in response to an increase in vehicular crashes, fatalities, and increased financial loss. To save costs, the Sacramento City Fire Department, Sacramento City Police Department, and the Sacramento County Sheriff's Department entered into a Joint Powers (Interagency) Agreement (JPA) for staffing and operating the facility.

Figure 2-1. Sacramento Regional Training Center skid pan.

Sacramento Metro Fire Department is a participating partner. The training facility is located on the former Mather AFB Alert Pad and was opened in mid-1999. The driving course was designed by the Emergency Vehicle Operations Course (EVOC) instructors and includes a skid pan that can handle sedans, ambulances, and engines (Figure 2-1).

Four full-time and two part-time fire instructors, six full-time and three part-time police instructors, and four full-time county sheriff instructors staff the facility. Fire personnel are cycled through the facility for annual training. Law enforcement personnel train on a 2-year cycle. The facility trains 900 to 1,100 fire personnel and 900 to 1,100 law enforcement personnel yearly. In addition, training is offered to any other government agency willing to enter into a cooperative agreement and pay a fee for the service. At the time of this writing, the facility costs approximately $3,000,000 a year to operate. The city contributed approximately $1.3 million and the county contributed approximately $1.7 million of the cost to maintain the facility.

Acknowledgements

The following are acknowledged for their expertise and willingness to give of their time in support of this project:

Bill Myers	Lieutenant, Sacramento County Sheriffs Office
Lloyd Ogan	Captain, Sacramento City Fire Department

SALT LAKE VALLEY (UTAH) AREA

Four agencies were interviewed in this area. The profiles of the fire departments and dispatch center are included here. The profile of the fourth agency, the Intelligent Transportation System (ITS), is included in the final profile of this section.

West Valley Fire Department

West Valley Fire Department serves a population of 114,000 and covers 42 square miles with 80 uniformed personnel. The department has five stations and six companies with the following distribution of apparatus:

3 ALS Type 1 engines	1 Raceway response unit
1 Tele-Squirt ALS engine	1 Decon trailer
1 Truck	1 Technical rescue squad
1 Haz mat squad	

In 2002, the department made 9,200 responses, with 82 percent being medical in nature. The department currently provides only ALS first response, but is in the process of assuming the entire EMS service.

Murray Fire Department

Murray Fire Department serves a population of 44,000 (with a daytime population of approximately 100,000) and covers 12 square miles with 50 uniformed personnel. The City of Murray recently annexed some of the surrounding area, so the department was in the process of conducting surveys to determine additional personnel/apparatus needs. The department has three stations with the following distribution of apparatus:

1 ALS squad	1 Haz mat squad
1 ALS engine	1 Utility
1 BLS engine	1 Brush truck
1 Truck	

In 2002, the department made 3,117 total responses, which are categorized as follows:

1,148 Fire
1,757 EMS
212 miscellaneous

This includes 324 traffic incidents and 17 auto/pedestrian responses. The department has an average response time of less than 2 minutes.

Valley Emergency Communications Center (VECC)

The Valley Emergency Communications Center (VECC) is a 36,000-square-foot, state-of-the-art facility that provides dispatch services to nine fire departments, eight police departments, public works, and animal control in the valley. The services cover a population of 3/4 million. The Center averages 4,000 calls per day, with 2,200 to 2,400 of those calls resulting in a dispatch of services. The Center is funded by a charge per call assessed to each department.

Figures 2-2 **a** and **b**. The VECC Fire Dispatch Pod.

The Center employs 3 managers, 10 supervisors, and 108 dispatchers. Based on call volume, 18 to 36 dispatchers are on duty at any given time. The Center is organized in three distinct dispatch pods — call takers, fire dispatchers, and police dispatchers (Figures 2-2 a and b).

Television monitors allow dispatchers to view certain traffic intersections through links to the Traffic Operations Center, Commuterlink cameras (Figure 2-3). The link provides views of intersections selected by Traffic Operations. The Center has no direct control of the cameras.

Figure 2-3. VECC Television Traffic Center.

Acknowledgements

The following are acknowledged for their expertise and willingness to give of their time in support of this project:

John Evans	Assistant Chief, West Valley Fire Department
Terry Hansen	Captain, HazMat
Russ Carr	Captain, Engine 75 Murray Fire Department
Randy Willden	Battalion Chief
Brad Freeman	Captain, Truck 81 Valley Emergency Communications Center (VECC)
Nancy McKean	Manager of Training and Professional Services

VENTURA COUNTY (CALIFORNIA) FIRE DEPARTMENT

In addition to the unincorporated areas, Ventura County Fire Department provides service to six cities within the county. The department covers 1,873 square miles, of which 860 square miles is a forest reserve. Service is provided to a total population of 452,584. The department has a complement of 549 personnel; 417 of those are sworn. There are 32 stations with the following distribution of apparatus:

50 Engines, Type I structural	15 Specialized response vehicles
10 Engines, Type III wildland	41 Fire crew/Construction vehicles
4 Trucks	21 Utility vehicles
2 Paramedic squads	113 Staff/Support fleet vehicles
9 Command vehicles	

Acknowledgements

The following are acknowledged for their expertise and willingness to give of their time in support of this project:

David Festerling	Deputy Chief
Mike Sandwick	Battalion Chief, Training
Kenny Kappen	Captain, Training
Terry Lamb	Firefighter, Training
Larry Jones	Fire Safety Risk Manager
Mark Karr	Captain
Larry Brister	Captain
Kevin Miller	Engine Driver
Michael Fine	Firefighter
Martin Gonzalez	Firefighter
Greg Chavez	Firefighter
Richard Snyder	General Manager, FAAC

VIRGINIA BEACH (VIRGINIA) FIRE AND RESCUE DEPARTMENT

The Virginia Beach Fire and Rescue Department is a career organization. In addition to fire suppression, the department provides basic life support (BLS) emergency medical care from engine, truck, and heavy rescue companies. The minimum certification for members assigned to these units is Emergency Medical Technician - Basic (EMT-B). The department also provides the opportunity for those holding ALS certifications to practice these skills. Each engine, truck, and heavy rescue company is assigned the necessary ALS equipment to fully function as the primary care provider for patients requiring advanced prehospital care.

The department has several specialty companies. The haz mat team is a partner in the Southside Regional Hazardous Materials response team. The technical rescue team is tasked with responding to emergency operations involving structural collapse incidents, trench and excavation collapse, confined space rescue, vehicle and machinery extrication, and high-angle rope rescue. The marine operations team maintains and operates a 21-foot vessel in the City of Virginia Beach and can respond within the port of Hampton Roads as requested. The team assists the Coast Guard and may be requested by any agency for assistance. The department also sponsors a FEMA US&R Task Force (Virginia Task Force II).

The department serves a population of approximately 439,257 people and covers 258.7 square miles of land area and 51.3 square miles of water area with a complement of 438 employees of which 413 are uniform personnel. There are 19 stations with the following distribution of apparatus:

21 Engines	11 Brush trucks
6 Trucks	2 Heavy rescue
4 Battalions	2 Tankers
	1 Trench rescue

The department averages 25,000 responses a year. The 2002 responses are categorized as follows:

Incidents Type	Number
Fire	8,704
EMS	15,648
Hazardous Materials	442
Other	2,491
False Alarms	1,694
TOTAL	28,979

Acknowledgements

The following are acknowledged for their expertise and willingness to give of their time in support of this project:

Virginia Beach Fire Department

David W. Land	Deputy Chief — Operations
John W. Harvey	Battalion Chief — Safety and Health
Patrick Cameron	Captain — B Shift Safety Officer
Steve Miles	Captain — A Shift Safety Officer
Captain Ray Irrizary and 7 fire fighters	Fire Station 8 — A Shift
Captain Pat Ehle and 7 fire fighters	Fire Station 9 — B Shift

City of Virginia Beach

Keith Barron	Risk Management Administrator
Ron Knowles	Engineer — Highways Department
Mike Shasiah	Engineer — Highways Department

INTELLIGENT TRANSPORTATION SYSTEMS

Two intelligent transportation system traffic centers were visited in the course of gathering information.

Commuterlink, Salt Lake City, Utah

The Utah Department of Transportation (UDOT) Commuterlink Traffic Operations Center (TOC) monitors and assesses the conditions of more than 85 miles of freeway system. The system currently uses 221 cameras mounted on 45-foot poles and spaced 1/2 mile apart. The Center can pan, tilt, and zoom the cameras. The cameras feed information across 250 miles of fiber optic cable with 13 hubs. The center also uses 61 variable message signs (VMS) and a highway advisory radio system to transmit current traffic information. Information from the cameras is fed to the control room's main screens (Figures 2-4 a and b).

Figures 2-4 a and b. The UDOT TOC console and main screen.

Congestion sensors are designed to detect traffic volume, speed, and congestion on freeways and surface streets in real time. Traffic sensors are spaced every 1/2 mile along the freeways. Weather recording stations and pavement icing sensors placed at key locations on the State highway system are used to alert motorists to snow-packed and icy conditions and to improve winter road maintenance and snow removal. The TOC is operational Monday through Friday, 5 a.m. to 11 p.m.; Saturday and Sunday, 10 a.m. to 6 p.m. Dispatchers are on call 7 days/week, 24 hours/day. The TOC provides a variety of traffic control, including:

- ramp meters on I-15 in Salt Lake and Davis Counties;
- variable message signs in Salt Lake and Davis Counties;
- portable trailer-mounted variable message signs;
- changeable directional signs;
- highway advisory radio;
- Incident Management Teams (IMT's); and
- distribution of traffic information via the Internet, live video feeds to local media, and emergency dispatch centers.

Two operators staff the Control Room on each shift, and a third operator is added during peak hours. Operators are responsible for the following:

- answer telephones;
- monitor and control views of traffic cameras;
- monitor police radio;
- monitor CAD data;
- monitor road conditions using sensor data;
- control VMS, traffic signals, and highway advisory radio;
- update Web site; and
- alert UDOT personnel and update UDOT road conditions hotline.

The IMT's cover approximately 140 miles over three interstates and a major highway traffic corridor. The teams operate from 6:30 a.m. to 7:00 p.m., 5 days a week, with personnel working 10-hour shifts. There are six IMT trucks: Three trucks are on the roads on Monday, six trucks operate Tuesday through Thursday, and three trucks operate on Friday. In addition to assisting at accident scenes, the IMT's perform a variety of motorist assist services, including tire changes/inflation, jump starts, adding water to a radiator, etc.

Acknowledgements

The following are acknowledged for their expertise and willingness to give of their time in support of this project:

David Kinnecom	Traffic Operations Engineer
Denny Simmons	Traffic Control Room Supervisor
Billy Frashure	Incident Management Specialist
Deryl Mayhew	Region Two Signal Engineer

Additional information about Utah's ITS, Commuterlink, and the TOC is available online at: http://www.commuterlink.utah.gov
or by contacting:
UDOT Traffic Operations Center
2010 South 2760
West Salt Lake City, Utah 84104

Smart Traffic Center, Hampton Roads, Virginia

The Virginia Department of Transportation (VDOT) Hampton Roads Smart Traffic Center monitors and assesses the conditions of 113 miles of interstate system using an advanced computer system linked to video cameras along the roadside and electronic sensors (Figure 2-5). The sensors, designed to detect deviations in traffic patterns, are linked by fiber optic cable to the center's computer system. The cameras allow the traffic controller to zoom in on traffic incidents from the center.

Figure 2-5. The Hampton Roads (Virginia) Smart Traffic Center.

The system currently uses 288 cameras mounted on 40-foot poles and spaced 1/2 mile apart. The cameras feed information across 522 miles of fiber optic cable to 70 TV screens in the control center. The center also uses 244 VMS and a highway advisory radio system to transmit current traffic information.

The center is operational 24 hours/day, 7 days/week. Controllers perform a variety of duties including:

- assess traffic patterns via closed-circuit video cameras;
- dispatch the Freeway Incident Response Team (FIRT) units to highway incidents;
- disseminate real-time traffic-related information via the Highway Advisory Radio and VMS;
- operate the I-64 HOV Reversible Roadway access gates;
- initiate and verify all VMS messages;
- support additional VDOT facilities (bridge openings, tunnel maintenance, etc);
- enter data and document information from daily operations via multiple software applications;
- share resources with VDOT facilities and bordering jurisdictions (including Transportation Emergency Operations Center);
- incident management support and coordination; and
- monitor radio communications between emergency responder agencies and the associated units.

The FIRT units patrol over 60 miles of interstate in the Hampton Roads area and perform a variety of motorist assist services, including tire changes/inflation, jump starts, adding water to a radiator, etc. Additional information on the Hampton Roads Smart Traffic Center is available online at: http://www.virginiadot.org/comtravel or can be requested via e-mail at hamptonroadsinfo@vdot.state.va.us. FIRT patrollers are dressed in highly visible apparel that meets ANSI standards (Figures 2-6 and 2-7).

Figure 2-6. FIRT vehicle.

Figure 2-7. FIRT patroller.

Acknowledgements

The following are acknowledged for their expertise and willingness to give of their time in support of this project:

Erika Ricks   Traffic Information Specialist
Ryan McLean   Lead Traffic Controller

APPARATUS SAFETY DEVICES

INTRODUCTION

This section reviews a variety of devices and markings that are placed on the apparatus. These markings and devices are used to improve the safety of firefighters riding in the apparatus, working at an emergency scene, and all those working in close proximity to the apparatus. Markings and lighting colors/patterns are used to increase the visibility of the apparatus. Some apparatus also incorporate devices that serve as reminders for personnel to follow safe practices when operating the apparatus. These devices range from simple markings to sophisticated technologies that monitor apparatus operation.

MARKINGS

The visibility of apparatus is a key component of responder safety. The following section examines the impact of vehicle striping and other markings.

Vehicle Striping

Retroreflective striping around the perimeter of fire apparatus is required in NFPA 1901, *Standard for Automotive Fire Apparatus.* The minimum requirements of the standard are intended to illuminate fire apparatus at night when visibility is limited. The retroreflective strip reflects the light produced by headlights and streetlights to provide an indication of the location and size of the apparatus. Recent additions to the standard require striping inside cab doors to help alert drivers to an open door.

The Phoenix, Arizona, Fire Department began using reflective striping in 1984. All vehicle striping installed on the front, sides, and back of the apparatus comply with NFPA 1901. The department implemented significant changes to its striping package in 1988 with the purchase of a large number of engines. Figures 3-1 a and b show both the NFPA-required white striping and an additional strip of red and white striping that is incorporated into rub rails at the base of the body. All apparatus lettering also is produced with retroreflective materials, adding to the visibility of the apparatus. The estimated cost of the additional striping and lettering was approximately $500 per unit.

Figures 3-1 a and b. Phoenix fire apparatus have NFPA-required striping and optional DOT-style, red/white truck reflective markings.

The white stripe above is required by NFPA 1901. The red and white stripe at the bottom of the body is supplemental retroreflective marking incorporated into the rub rail.

The pattern of the additional red and white striping along the rub rails is that required by NHTSA for commercial trailers and truck tractors and is often referred to as "DOT conspicuity striping." This material is widely available and inexpensive. Although adding some expense, incorporating the striping into the vehicle's rub rails adds protection for the striping material. More information on the conspicuity requirements for commercial trucks and tractors can be found online at: http://www.fmcsa.dot.gov/pdfs/consp.pdf

The Plano, Texas, Fire Department marks apparatus with conspicuous, contrasting colors to improve driver recognition while the apparatus is on the roadways, both moving and stopped. While a great deal of attention and money is spent to make the front of fire apparatus conspicuous, many fire departments pay very little attention to the rear of the apparatus. Plano's current vehicle marking system has been in effect since approximately 1998. It is based on markings that the fire chief observed on the rear of fire and law enforcement vehicles while in the United Kingdom. Vehicle markings are reviewed regularly to determine if the pattern can be redesigned to be more noticeable.

The rear of the apparatus has a barricade pattern with 6 inch red stripes on a lime-yellow background (**Figures 3-2 a** through c). The lime-yellow sheets go on the smooth surfaces of the rear of the apparatus. The 6 inch red stripes go over the lime-yellow background. The cost of the herringbone pattern on the rear of apparatus was approximately $1,000/unit.

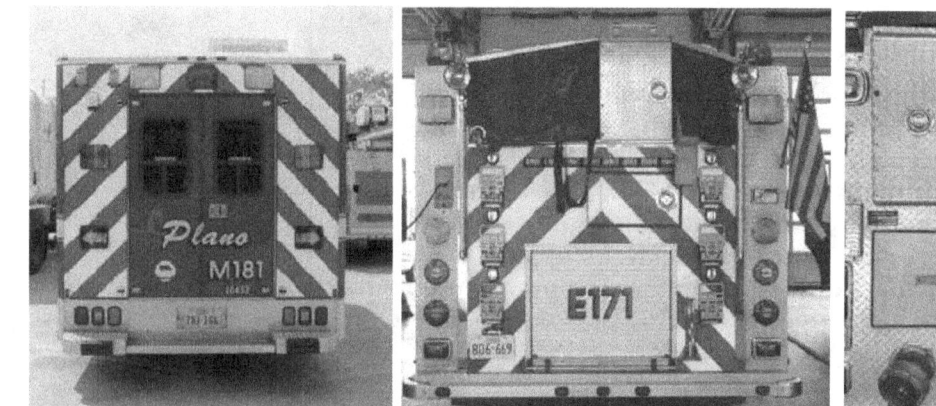

Figures 3-2 a through c. Plano apparatus feature U.K.-style reflective patterns on the rear of the vehicle.

The specifications for the barricade pattern are
- Lime-yellow background material: 3M Scotchlite® Diamond Grade Fluorescent LDP Reflective Sheeting — 3963 — Yellow Green.
- Red 6-inch diagonal stripes: Scotchlite® Electronic Cuttable Film — 1172 — Red.

Optional treatments for the rear barricade pattern:
- 1160 Protective Overlay Film. (This is like putting Teflon over the reflective film, making it easier to clean.)
- 880i (ink) clear edge seal. (This keeps the dirt out of the first row of cells at the edge of the sheeting.)

The striping reflects brightly off lights, chrome, etc., and is very visible to the traffic that faces the rear of apparatus, especially at night. Personnel provided a specific example of the improved vehicle visibility: While they were driving to another department for a night training exercise, tractor-trailer drivers commented on the citizen's band radio about the brightness of the design. The visibility of the rear marking is decreased at incident scenes when an apparatus is angled to provide blocking protection for emergency workers.

Although not used by any of the departments that were visited in the preparation of this report, fluorescent apparatus paint is also available. This paint adds to the visibility of the apparatus in daylight and at night. The paint tends to be more expensive than standard apparatus finishes.

Other Markings

Phoenix fire apparatus have several reminder markings for safety. There is a reminder posted on the windshield above the driver to use a backer while in reverse (Figure 3-3). There are also height reminders posted on both the mirror and windshield (Figure 3-4) and an internally mounted "DO NOT MOVE" light (Figure 3-5). The "DO NOT MOVE" light is activated if any cab or compartment door is open. This light is required by NFPA 1901. An audible buzzer is another way to alert apparatus drivers and officers of an open cab or compartment door. The buzzer should have a cutoff feature that disables it when the vehicle's parking brake is set. This will avoid excessive background noise if the Incident Commander operates from inside the cab.

Figure. 3-3. This reminder to have spotters is used on Phoenix fire apparatus.

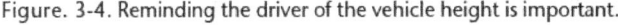
Figure. 3-4. Reminding the driver of the vehicle height is important.

Figure. 3-5. The "DO NOT MOVE" light on a Phoenix engine company.

RESTRAINTS

NFPA 1500, *Standard Fire Department Occupational Safety and Health Program*, specifies the mandatory use of restraints during both emergency and nonemergency response. It can be difficult for company officers to assure all responders are in compliance when the restraints are a neutral color that blends in against apparel. Plano, Fire Department apparatus have orange-colored seat restraints to make it easier for the officer and others to assure that all personnel are restrained before the apparatus moves. The rearview mirror above the officer's seat allows him/her to assure that all personnel are wearing restraints without the distraction of having to turn around (Figures 3-6 a and b).

A recent change to NFPA 1901 requires red seatbelt webbing (section 14.1.3.1). This change was based on a public comment received by the NFPA apparatus committee from participants of the EVSI forums. After review, the committee voted to include the requirement in the new edition of the standard. All fire apparatus built to NFPA 1901 will incorporate this safety feature.

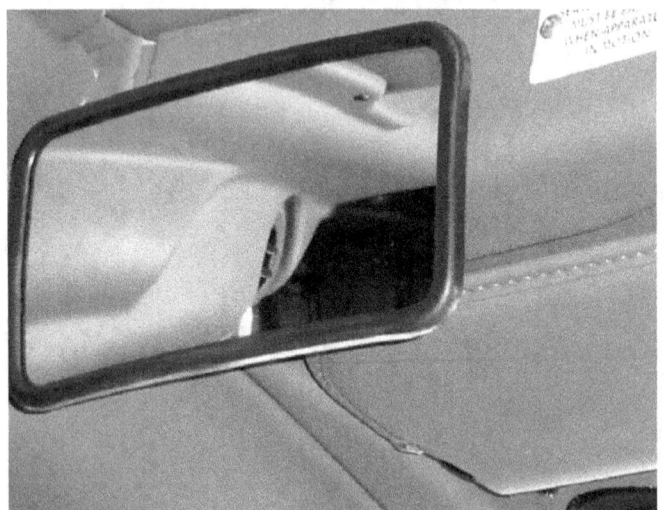 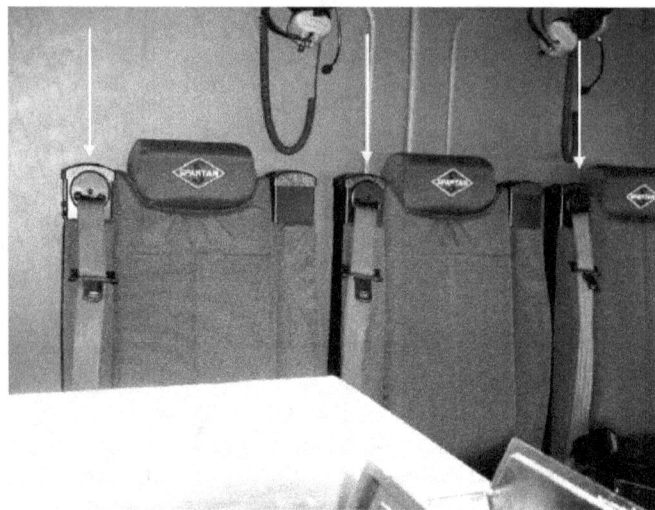

Figures 3-6 a and b. Orange restraints and the officer's seat rearview mirror on Plano fire apparatus.

WARNING LIGHTS

A system of optical warning devices (warning lights) is required by NFPA 1901 for all fire apparatus. The standard addresses emergency lighting in the upper and lower zones of the apparatus and the requirements for all four sides. The standard segments the required emergency lighting into two modes.

The 'calling for right of way' mode describes the emergency lights that operate as the vehicle is in motion or responding to the scene of the emergency. These lights are designed to catch the attention of other drivers and pedestrians, give them notice that the fire apparatus is approaching, and allow them time to yield to the apparatus. The 'blocking right of way' mode describes the lights that operate when the fire apparatus is parked on the scene of the emergency. This lighting is intended to alert drivers that the vehicle is stopped and allow the drivers to take action to avoid colliding with the parked emergency vehicle.

NFPA 1901 allows the use of red and blue emergency lights in both modes. Clear (white) lights are not allowed to project toward the rear of the apparatus while in the 'calling for right of way' mode and are not allowed to operate at all when the vehicle is in the 'blocking right of way' mode. Clear lights to the rear of the apparatus can blind drivers behind the emergency vehicle as it is in motion, and blind drivers approaching from any direction if used on the scene of an emergency. Amber (yellow) lights are allowed to project only forward when the vehicle is in the 'calling for right of way' mode, but may project in any direction while the vehicle is in the 'blocking right of way' mode.

Authorization to operate different colors of warning lights varies by State and different light colors may have alternate designations. For example, blue lights signify a volunteer firefighter in New York State, while blue lights represent law enforcement, EMS, and fire vehicles in other parts of the country.

The most effective lighting continues to be debated. What is known is that visibility, conspicuity, driver attention, and driver expectancy are the factors that determine whether an object is perceived appropriately.[1] Visibility and conspicuity must be considered in the design of warning lights because these factors are not highly dependent on driver behavior.[2]

The sensitivity of human vision peaks in the yellow-green portion of the spectrum. It has been established that white is the most visible color for warning lights, followed by green, amber, and red.[3] White alone fails to identify an emergency vehicle. Although green is visually effective, it is viewed as a "go," or "safe," color in our society.[4] Yellow and red signify "danger," which led to their popularity as warning and caution identifiers.

The California Highway Patrol (CHP) conducted studies in 1971 and 1974 to determine if vehicle lighting factors were responsible for an increase in rear-end collisions. Their data was inconclusive and did not include side-swipe incidents.[5] However, the studies did stimulate further research by Stafford, Rockwell, and Banasik[6] in 1970 and Jan Berkhout[7] in 1979. Both follow-on studies showed the color blue induced an illusion of approaching motion and the color red induced an illusion of receding motion.

Plano Fire Department uses additional lighting of a contrasting color to the vehicles. Figures 3-7 a and b show the department's red EMS units with two contrasting blue strobes on the light bar and one blue strobe under each headlight to contrast the red color of the vehicle.

Figures 3-7 a and b. Contrasting blue lights on Plano ambulances.

In 1994, a Phoenix Fire Department firefighter was killed when struck by a passing vehicle. Following the fatality, the department studied onscene emergency lighting. Findings suggested that amber (yellow) lighting was less likely to blind drivers and also less likely to draw the interest and attention of passing drivers. It was thought that approaching drivers (seeing amber lights rather than red, clear, or blue) would believe that they were approaching a highway or tow truck rather than an emergency scene and take action to avoid the hazard. It was also believed that drivers seeing red, clear, or blue lights were more likely to take an interest in examining the scene as they passed. This could divert their attention long enough to cause a crash or could actually draw drivers to the flashing lights.

At the time of this writing, 10 engines were configured for all nonamber warning lights (clear, red, and blue) to go off when the apparatus parking brake was engaged. Amber lights on all four sides of the apparatus are the only functioning lights in the 'blocking right of way' mode. The amber lights are not tied to directional arrows. Figures 3-8 a and b (on page 22) show the amber lighting configuration on the front and the rotating amber light on the side of the engine.

Figures 3-8 a and b. Amber lighting used on Phoenix fire apparatus.

Figure 3-9. The amber lighting override.

The apparatus is equipped with an override switch that allows the driver to continue using the warning lights, except clear flashers, even when the parking brake is engaged (Figure 3-9). The estimated cost of installing this system was $200 per unit. The estimated amperage drawn on apparatus when only amber lights are operating is 30 amps, as opposed to 100 amps when the override is activated and 130 amps when the unit is responding with lights and siren.

According to maintenance personnel, the additional relays, overrides, and interlocks on warning lights have had minimal to no effect on apparatus maintenance frequency or severity.

Personnel interviewed believed that many companies routinely used the override at scenes. Interviews with line personnel revealed that some used the override routinely during daytime operations but not at night, believing the amber lights alone do not provide the necessary visibility during daylight hours. No personnel interviewed noticed a difference in traffic reaction to all amber lights when operating on scenes.

ELECTRONIC MONITORING EQUIPMENT

Modern technology has provided the fire service with a variety of electronic devices that can assist in improving fire apparatus and firefighter safety. The following section highlights some of the devices that are installed on modern fire apparatus.

Backing Cameras

Backing cameras are installed on all Murray, Utah, Fire Department engines and trucks. The cost of the cameras was approximately $475 each, uninstalled. These cameras are mounted on the rear of the apparatus with the screen mounted next to the driver (Figures 3-10 a and b). Figure 3-11 shows the view provided by the backup camera. The apparatus is parked in the bay with the rear doors open.

Figures 3-10 a and b. Backing cameras are installed on the rear of the apparatus.

Figure 3-11. The cab-mounted monitor allows the driver to view the area behind the apparatus.

Even though cameras provide an additional measure of safety, they do not substitute for spotters. NFPA 1500 requires spotters for backing, regardless of whether the apparatus is equipped with cameras or other backing safety equipment.

Recording Cameras

The City of Plano installed the DriveCam® in 911 city vehicles, including fire department apparatus (Figure 3-12). The DriveCam® is a palm-sized video event data recorder that mounts inside the vehicle behind the rearview mirror and monitors driving activity continuously. The cost was approximately $1,000 per unit, uninstalled.

The unit monitors video, audio, and four directional G-forces caused by activities such as hard braking, sudden acceleration, hard cornering, and/or collisions. Data collected through monitoring are purged on a first-in, first-out sequence, unless the device is triggered to record. When any directional G-force exceeds the preset threshold, the event is recorded for later viewing. A recorded event contains the data from 10 seconds prior and 10 seconds after recording is triggered. The driver can also activate the unit manually at any time. The unit can store 17 to 24 incidents in its flash memory. Incidents are downloaded from the camera to a computer for viewing.

The city plans to use the data from the units to review accidents involving city vehicles, as well as for legal purposes. Although the specific data analysis has not begun, the city Risk Management department is currently working to incorporate the information gathered from the DriveCam® into their RiskManager® software package in an effort to better track losses from vehicular-related incidents.

Figure 3-12. The DriveCam® that is installed on Plano fire apparatus. *Photo courtesy of Chereskin Communications*

On-Board Computers

Another electronic device used to monitor the safety of drivers is the on-board computer, commonly called the Black Box. The Black Box monitors parameters including speed, rpm, vehicle stops, excess idle time, overspeeds, over rpm's, hard accelerations, hard decelerations and daily distance. If specified by an organization, the system can also monitor the use of restraints, turn signals, brakes, emergency lights, siren, auxiliary power, doors, and warning lamps. It provides feedback to the driver when unsafe or aggressive actions are taken and keeps a computer log of driver performance.

When the driver begins to operate the vehicle in an unsafe manner, the system produces an audible warning tone, much like a Geiger counter. If the warning is ignored, the system produces a steady tone. This tone alerts the driver that the system has started recording the occurrence. The computer log information can be downloaded to an office computer for analysis or has an option to transmit data wirelessly to a central data collection point. The program also contains preformatted reports. At the time of this writing, the cost of the system was approximately $2,500.

RECOMMENDATIONS FOR APPARATUS SAFETY DEVICES

- Mark apparatus with conspicuous, contrasting colors.
- Consider visibility and conspicuity when designing color and placement of additional warning lights on vehicles.
- Consider contrasting colored restraints and a rearview mirror above the officer's seat.
- Use spotters when backing the apparatus.

[1] Hills, B.L. (1980). as cited in DeLorenzo, R. & Eilers, M. (1991). Lights and siren: A review of emergency vehicle warning systems. *Annals of Emergency Medicine*, online.

[2] DeLorenzo, R. & Eilers, M. (1991). Lights and siren: A review of emergency vehicle warning systems. *Annals of Emergency Medicine*, online.

[3] Allen, M., Strickland, J., & Adams, A. (1967). as cited in DeLorenzo, R. & Eilers, M., op. cit.

[4] Scarano, S. (1981). as cited in DeLorenzo, R. & Eilers, M. (1991), op. cit.

[5] Davis, C.C. (1982). *Accidents involving stopped vehicles on freeway shoulders.* Automobile Club of Southern California.

[6] Stafford, R.R., Rockwell, T.H., & Banasik, C.R. (1970) as cited in Davis, C.C., op. cit.

[7] Hills, B.L. (1980). as cited in DeLorenzo, R. & Eilers, M. (1991), op. cit.

TRAFFIC CONTROL MEASURES

INTRODUCTION

Control and navigation of busy intersections is a dangerous situation during a response. Numerous studies have shown that intersections are the most likely location for an emergency vehicle to be involved in a crash. Managing traffic flow during highway operations also is critical to the safety of responders. This section reviews the use of optical preemption for safely navigating controlled intersections and the use of State DOT ITS and highway response teams to assist in traffic management during highway operations. Although this discussion focuses on Opticom®, the brand used by the departments visited, the reader should be aware there are several other manufacturers of preemption devices.

OPTICAL PREEMPTION

Optical preemption is one method of improving navigation through intersections when responding in the emergency mode. The safety of both civilian drivers and responders is improved when emergency vehicles cross through an intersection on a green light. Responders can control signals with optical preemption, resulting in a green light when the emergency vehicle arrives at the intersection. Initiating optical preemption usually provides adequate time for civilian traffic to clear the intersection before an emergency vehicle arrives.

The majority of the fire departments in the Salt Lake Valley have optical preemption devices installed on at least some major intersection traffic lights. The Murray Fire Department was the first to install the devices, and has used optical preemption for over 8 years. At the time of this writing, 75 percent of the city's major intersections were covered with optical preemption. Optical preemption installation is required now on all new controlled intersections built in the city.

Response time was not a factor for installing optical preemption. The decision to install the devices and the choice of brand (3M Opticom®) used in Murray was made by a previous chief who came from an area that had optical preemption. The department budgets for the installation of the device on one major existing intersection per year. The cost to the department for 2003 was $5,500. Other brands were reviewed, but there was a question of compatibility with existing units.

UDOT approves the installation of traffic signal preemption by Salt Lake Valley area fire departments and requires a signed agreement from each department. The agreement requires the city to pay for installation, and UDOT agrees to maintain the signal devices. Some cities now require developers to pay for installation of preemption as the traffic signals are installed. UDOT must also approve the vendor. This requirement was implemented to assure that preemption devices are able to identify and code units and to prevent vendor monopoly. Although one goal is to prevent vendor monopoly, the engineers agree that followup on compliance has not been a priority. The UDOT agreement form can be found in Appendix B.

The decision to install the devices in Virginia Beach was based on increasing traffic volume resulting in traffic jams at key intersections, an increase in the number of blind intersections, and an increase in the number of near-misses for emergency vehicles. In 1993, the Traffic Engineering Division of the Highways Department made the decision to use 3M Opticom®. It was selected because:

- It had been tested and used by other local governments.
- Traffic Engineering can control the system. The system identifies each vehicle that uses the device, including date and time. This prevents unauthorized use.
- The system is customized for use by fire and EMS only.

The Traffic Engineering Division maintains 330 controlled intersections in Virginia Beach. As of June 2003, the preemption device was installed on 45 of those intersections. At the time of this writing, the cost of installation was $7,500 per intersection.

Opticom® has two settings, high and low. Both areas have it set on high, which activates the traffic signal from 1/2 mile away and holds it on green for 15 seconds after it loses the preemption signal. It takes approximately 12 additional minutes to recycle the signal back to the ITS setting. The device does have limitations. The emitter functions on line-of-sight (Figure 4-1). Any obstruction to that line of sight, such as signals that are around curves or are blocked by trees in a median, results in a failure to activate the signal change until they are cleared. Fog and snow also can decrease the distance of activation.

Figure 4-1. The Opticom® Priority Control System line-of-sight. *Courtesy of 3M Manufacturing*

All Virginia Beach Fire and Rescue marked vehicles have the preemption device installed. Installation of the device on the light bar is part of the specification and bid process for all new apparatus and has significantly increased the cost (Figure 4-2a). Installation includes an automatic control that activates the device when the warning lights are turned on and deactivates the device when the parking brake is applied. Reserve units are being upgraded as older units are retired.

All Murray Fire Department vehicles with light bars have the device (Figure 4-2b). Although the device is deactivated automatically when the lights are shut off or the parking brake is engaged on new apparatus, the department has experienced a prob-

Figure 4-2a. This emitter is mounted within the warning light bar.

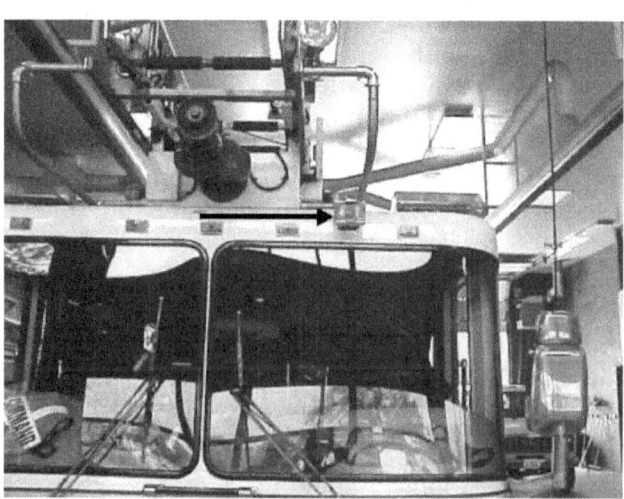

Figure 4-2b. This emitter is independent of the vehicle warning light bar.

lem with personnel failing to manually deactivate the signal on some older apparatus. Emission control in the older apparatus is tied to door opening or switches. Failure to manually deactivate has resulted in holding the light at green and tying up traffic for a considerable distance. It has often taken someone else to notice the traffic back up and call the apparatus to tell them to deactivate the device.

Training Personnel

The manufacturer provided the initial training for personnel when the device was first installed on the Murray Fire Department vehicles. The department then set up driver training to practice device activation and observation of results. The initial installation brought calls from citizens questioning what was happening. The department sent an information flyer in the public works bill to explain the new system to all citizens. When the device was installed on Virginia Beach Fire and Rescue vehicles, guidelines for operation were sent to personnel in a departmental directive that was passed from the chief down through the ranks. Training was completed in-station by the captains.

Effects on Traffic

The use of optical preemption devices has notable affects on intersection control by approaching emergency vehicles and the response by civilian motorists. Some of these are highlighted as follows.

INTERSECTION CONTROL

A major problem can arise if apparatus from opposing directions attempt to control the signal at the same time. In 1990, the Plano, Texas, Fire Department had a serious crash between department apparatus when an engine proceeded against the red light after a ladder truck already had the preemption. The engine driver was so used to capturing the light that he assumed he had captured the light when he had not.

Both Virginia Beach and Murray department personnel cited occasional instances of opposing vehicles attempting to control the signal. Since drivers understand that the first unit to activate the preemption device is the unit that controls the intersection, this has never resulted in any mishaps. Both departments' policies require units to stop at red lights before proceeding.

CIVILIAN DRIVER RESPONSE

Even though both departments activate the unit from the same 1/2-mile distance, descriptions of the preemption's effect on civilian driver behavior by department personnel were almost opposite. Murray Fire Department line personnel stated the preemption usually clears the intersection before the unit arrives and allows it to keep rolling at a safe and relatively steady speed. Personnel have had no problems with drivers appearing confused or not clearing the intersection. Preemption also has eliminated the need for units to jump medians to get around intersection traffic, which reduced replacement of chassis springs.

Line personnel from Virginia Beach described a variety of motorist responses to signal preemption, including:
- appears visibly shocked/surprised;
- slams on brakes;
- displays visible anger if more than one unit preempts the light while they are stopped;
- unaware that the light has been preempted; and
- proceeds through the preempted light anyway — this occurs more frequently at the hard-wired preemption in front of the fire stations.

Although these behaviors were noted, there have been no intersection crashes since the system was implemented.

Data Analysis and Findings

Neither Virginia Beach nor Murray conducts specific data analysis, such as response times or intersection crashes, related to optical preemption. The Virginia Beach Risk Management department does track all automobile and property claims against police, fire, EMS, and all other agencies operating for the city. A summary of the automobile liability for the fire department for 5 years is given in Table 4-1.

TABLE 4-1. SUMMARY OF FIRE DEPARTMENT-INCURRED CLAIMS, AUTOMOBILE LIABILITY VIRGINIA BEACH, VA		
Fiscal Year	Claim Count	Incurred
1998	30	$ 8,962.97
1999	9	$209,073.00
2000	19	$ 5,442.06
2001	15	$ 15,360.77
2002	18	$ 87,139.33
TOTAL	91	$325,978.13

Subjectively, line personnel from both departments agreed that the use of preemption provides improved and safer intersection navigation by opening a clear path for the emergency vehicle to pass through and reduces response times. Murray personnel cite the time it took to run the length of the city's main street as an example of improved response times. During the peak Christmas shopping rush before preemption, it took approximately 15 to 18 minutes. After the installation of preemption devices, the time decreased to approximately 5 to 7 minutes. They also suggest that the department performs less maintenance, especially on brakes, because units don't have to consistently start and stop.

Before it was installed, the UDOT traffic engineers believed that optical preemption would create significant traffic problems. They now agree that preemption does not really hurt traffic flow and improving technology is making it even better. Although there is currently no way to track the number of preemptions remotely, that information can be manually downloaded from a given signal, if there is a complaint or reason to do so. The traffic engineers suggest yearly tests to be sure the devices are working. This test could be done in conjunction with the annual signal inspection done by the engineers.

INTELLIGENT TRANSPORTATION SYSTEM, TRAFFIC CONTROL CENTERS

An ITS and a TOC can provide considerable assistance to emergency response departments for traffic control at a highway operations scene. The FIRT units in Virginia Beach are dispatched to assist at highway incidents by the Hampton Roads Smart Traffic control center. The IMT units in the Salt Lake City area must be requested and dispatched by the State police to assist at highway incidents. The IMT's monitor fire frequencies and, at the time of the site visit, were working to get permission to talk directly to fire units enroute.

All FIRT and IMT vehicles carry traffic cones and are equipped with a lighted arrow board to control the traffic flow. The vehicles have an amber strobe light located on top of the cab. When the arrow board is deployed, it blocks the amber strobe, reducing confusion and allowing motorists to focus on the arrow board. The board can be raised or lowered and the desired arrow pattern activated from within the cab for added safety. Figures 4-3 a and b show the lighted arrow board on a Utah IMT vehicle.

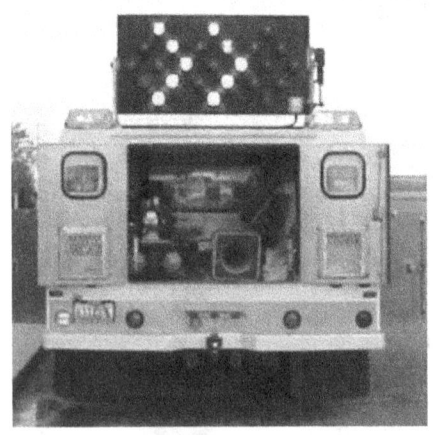

Figures 4-3 a and b. A Utah DOT IMT vehicle arrow board.

UDOT IMT vehicles are equipped with vertically extendable floodlights on both the driver and passenger side of the vehicle (Figure 4-4). These also are controlled from inside the vehicle.

UDOT IMT vehicles remain on scene as long as necessary. If it is determined that an incident will exceed 1 hour, the FIRT patroller contacts VDOT to obtain larger equipment including, but not limited to:

- arrow trailers;
- additional cones;
- temporary signage; and
- VDOT trucks with rear-mounted bumper blocks and hatch marks.

Departments that do not already have a working relationship should contact their State's DOT to identify criteria for responses that would incorporate DOT resources to aid in traffic control and improve safety for responders involved in highway operations.

Figure 4-4. A Utah DOT IMT extendable floodlight.

RECOMMENDATIONS FOR TRAFFIC CONTROL MEASURES

- Consider optical preemption to improve emergency vehicle movement through intersections.
- Require units to come to a complete stop at red lights, stop signs, and activated or unguarded rail crossings before proceeding.
- Pursue a working relationship with the State DOT to identify criteria for responses that would incorporate DOT resources to aid in traffic control and improve safety for responders involved in highway operations.

HIGHWAY OPERATIONS

INTRODUCTION

Our society is mobile. Every year the volume of traffic increases and interstate systems become more complex. We spend hours traveling to and from work in urban areas. In response, automobile manufacturers make vehicles as comfortable as possible. They control noise and provide a variety of entertainment, including radios, CD's, and screens for watching videos.

We also are a society in a hurry. Because we must spend more time on the roadways, we conduct many activities while traveling from point A to point B. Studies have shown that these activities include eating, putting on make-up, combing hair, reading the paper, and the ever-present talking on the cell phone. All these activities decrease focus on driving and traffic flow.

We are curious when seeing emergency lights along the roadside. Drivers routinely try to see what is going on, often missing an altered traffic pattern. Any impairment from fatigue, alcohol, or drugs increases the potential for poor judgment and a mishap.

Highway operations should be handled in the same manner as other types of operations. Departments must preplan highway emergency incidents with emphasis on tactical considerations and personnel safety. This section will review practices that provide a safer environment and improve the protection of emergency responders involved in highway operations.

APPARATUS PLACEMENT

The Phoenix, Arizona, Fire Department made major revisions to their positioning SOP after the loss of a firefighter in 1994. The firefighter/paramedic was struck by a drunk driver while loading a patient into the back of an ambulance. Prior to this fatality, the use of an emergency vehicle as a barrier to protect onstreet incidents was well established. The SOP was used for highway and freeway emergencies, but not for incidents in residential areas. The positioning SOP was changed to place emphasis on the dangers in residential areas, as well as on highways. Personnel interviewed for this report stated that the level of awareness of apparatus positioning is significantly higher than before the 1994 fatality.

A line-of-duty fatality that occurred in Midwest City, Oklahoma, prompted the Plano, Texas, Fire Department to institute an apparatus positioning policy. On August 5, 1999, two career firefighters were struck by a motor vehicle on a wet and busy interstate. One firefighter was killed and one was seriously injured. The NIOSH report of this incident is available online at:

http://www.cdc.gov/niosh/summ9927.html

At the time, there was no consistency for securing a highway incident scene in Plano. Although the department had not suffered a catastrophic loss, they realized the potential and the need to develop a highway safety operations policy. As in other growing urban/suburban areas, increases in traffic volume and speed also contributed to the need for improved safety on the roadways. The Plano policy mirrors the policy developed by Phoenix Fire Department.

Fairfax County, Virginia, Fire and Rescue developed a manual to outline appropriate highway operations. This manual, *Operating Procedures for Highway Incidents*, is now used in place of a SOP. The manual was developed in response to an increase in the incidence of near-misses while operating on the highway. It is based on information from Federal DOT publications and was developed in consultation with VDOT.

The SOP's of all three departments provide graphic examples of vehicle placement. Figures 5-1 through 5-8 are the Phoenix apparatus placement graphics.

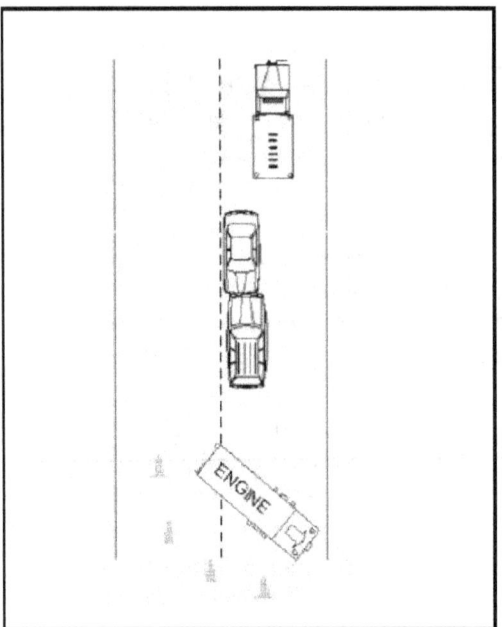

Figure 5-1. Where possible, angle apparatus at a 45-degree angle from the curb.

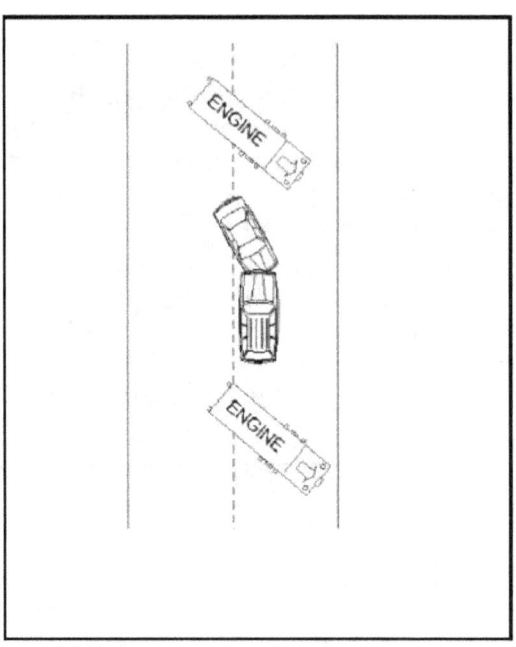

Figure 5-2. Prioritize placement of the apparatus by blocking from the most critical to the least critical side.

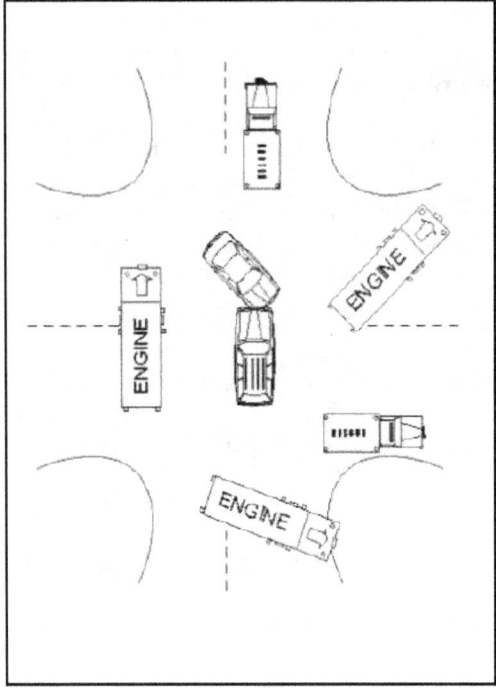

Figure 5-3. Often times two or more sides may need to be protected.

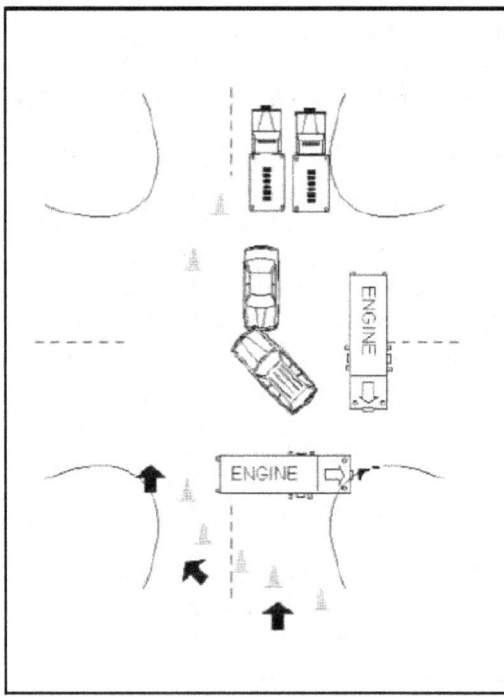

Figure 5-4. Traffic routes around the fire apparatus.

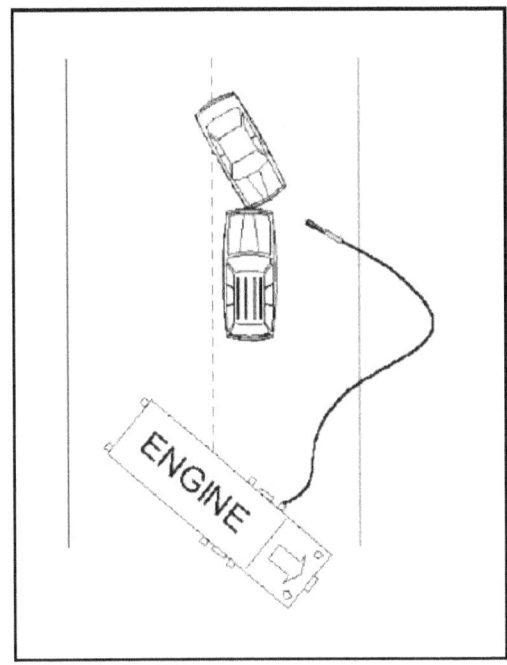

Figure 5-5. To protect pump operator, position apparatus with the pump panel on the side opposite on-coming traffic.

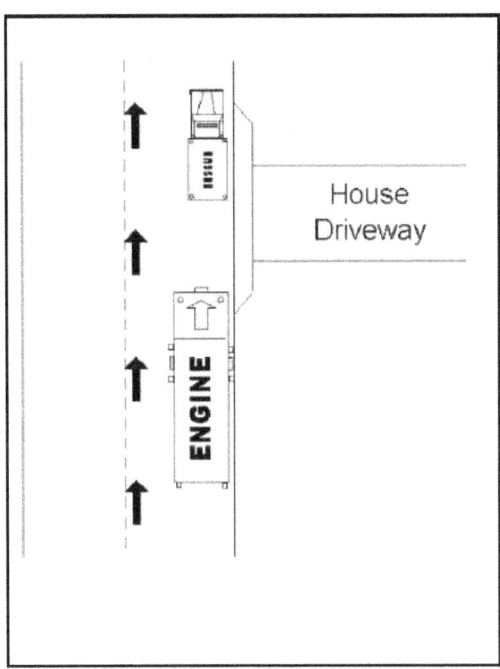

Figure 5-6. Where possible, park rescues in driveways or position rescue to protect patient loading area.

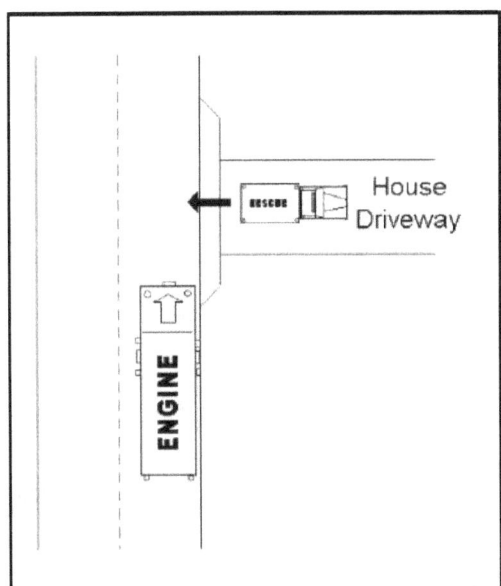

Figure 5-7. Protect the ambulance loading area even when the ambulance is parked in a driveway.

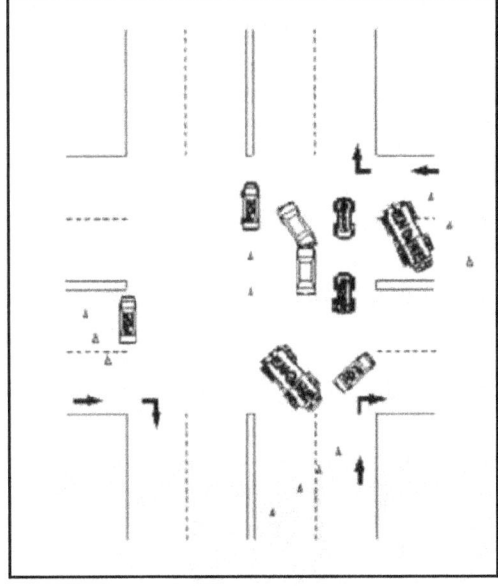

Figure 5-8. Provide specific direction to police as to what traffic control needs you have. Position rescues to protect patient loading areas.

Training

When the apparatus positioning SOP was changed in Phoenix, much of the training was done using videos that were aired on the cable television Phoenix Fire Network (PFN). PFN is a 24-hour-a-day operation, and schedules programs based on input from Operations personnel and the Safety Officer. Practical training was done at the district level, using demonstrations of how to position, how to set up work zones, and some simulated scenes for hands-on practice.

The Plano training included classroom presentation of the Midwest City incident. Videos, pictures, and other information regarding that incident served as the basis for the safety discussion. A tabletop exercise planned the placement of apparatus at various intersections and locations on various roadways. This was followed by a road-way exercise with apparatus to emphasize the points of the classroom and tabletop exercise. Personnel consistently remarked that the concept was discussed so often in advance of the policy and training, it seemed like second nature when it went into effect.

Fairfax County recruits receive the highway operations manual while at the academy. Other personnel are given the manual when they go through driver training. Policy change is communicated via the Intranet. The change may be in the form of a general order, a training or safety bulletin, or a manual change. This is usually followed by an in-station discussion and/or training, or may involve quarterly training at the academy.

SCENE LIGHTING AND MARKING

Lighting specifications for new apparatus for all departments interviewed are based on NFPA 1901. Fairfax County Fire and Rescue personnel cite an increased use of amber lighting in conjunction with emergency lighting, but amber lights are not used in isolation. The current policy is to leave all emergency lights on at the scene. Headlights are turned off when facing oncoming traffic. Figures 5-9 through 5-11 show different configurations for amber directional arrow placement on Fairfax County fire apparatus. Figure 5-12 depicts cone placement for traffic control and apparatus placed in the "fend off" (45-degree angle) position to protect the work zone and emergency providers. Figure 5-13 depicts night flare placement.

Plano Fire Department uses the directional arrows on the rear light bar of apparatus, turns off headlights, and leaves rotators and flashing beacons on while operating at highway incidents. When floodlights are used, they are raised to a height that allows light to be directed down on the scene. Personnel believe this provides the optimum working conditions at night, improves vision, and reduces the trip hazard

Figure 5-9. A large directional arrow on the rear of a Fairfax County engine company.

Figure 5-10. Some Fairfax County apparatus are equipped with arrowstick light bars.

Figure 5-11. A combination of apparatus-mounted warning arrows and traffic cones should be used to redirect traffic.

Figure 5-12. Apparatus should be angled at 45 degrees, and cones placed to route traffic around the scene.

Figure 5-13. Flares are more visible than traffic cones during nighttime operations.

by reducing shadows. "The more lights, the safer the providers" is the consensus among Plano line personnel. The Plano Police Department also has a support pickup truck with cones that can be called to help divert traffic from an incident scene.

The Virginia Beach Fire and Rescue SOP for highway markings was developed as the result of proactively reviewing safety aspects of highway operations and an increasing number of responses for highway incidents. Line personnel also identified the number of near-misses they experienced and injuries in the neighboring Norfolk Fire Department as factors that prompted the desire for the SOP at their level.

Virginia Beach used a team approach to develop the SOP, which is based on the Fairfax County and Phoenix departments' policies, NFPA standards, State regulations, and what Fort Collins, Colorado, was doing related to lighting issues. The department also consulted public safety, law enforcement, VDOT, and the private EMS provider to identify their needs and recommendations.

The policy for highway marking distances for a working incident zone (300 feet when light, 500 feet when dark) was based on vision and distance changes at night. The policy also identifies the minimum requirements for the work zone. The Incident Commander can adjust these distances, request additional units for protection, and block the entire road as dictated by existing conditions. The State police are responsible for traffic control outside the fire department's established work zone.

The "best" light color(s) and lighting pattern for highway operations is still being debated. Research by the National Institute of Standards and Technology has shown that as the number of flashing lights increase, the ability of drivers to quickly respond to the emergency message decreases.[1] A 1988 study conducted by Boff and Lincoln at Wright-Patterson AFB showed an emergency flashing light is noticed quickly if there are no other flashing lights in the field of view.[2]

Dr. Stephen Solomon, an ophthalmologist who has studied emergency vehicle colors and lighting, notes that the fire service philosophy has been to attract as much attention as possible through a combination of lights and light colors with varying degrees of reflection and flashes. Strong stimuli hold central gaze and drivers tend to steer in the direction of gaze. If fatigue, alcohol, or drugs impair the driver, the potential and degree of drift increases. He suggests this practice actually makes the fire apparatus a "visual, magnetic target." He recommends reducing the time span of looking toward a complex flashing light display by reducing the number, brightness and array of color, revolving strobe, and reflecting lights during emergency travel;

and using either filament bulbs in one or two amber flashers (mounted on the upper level of the vehicle on each corner) blinking in tandem or revolving beacons when the vehicle is parked along the road or at a curb and clear of all active traffic lanes.[3]

The use of emergency vehicle lighting is also addressed in the Manual of Uniform Traffic Control Devices (MUTCD). The manual revision had a target finalization date of October 2003. The manual addresses the use of emergency lighting as follows:

Section 6I.05 Use of Emergency-Vehicle Lighting (Flashing or Rotating Beacons or Strobes)
Support:
The use of emergency-vehicle lighting is essential, especially in the initial stages of a traffic incident, for the safety of emergency responders and persons involved in the traffic incident, as well as road users approaching the traffic incident. Emergency-vehicle lighting, however, provides warning only and provides no effective traffic control. It is often confusing to road users, especially at night. Road users approaching the traffic incident from the opposite direction on a divided facility are often distracted by emergency-vehicle lighting and slow their vehicles to look at the traffic incident posing a hazard to themselves and others traveling in their direction.

The use of emergency-vehicle lighting can be reduced if good traffic control has been established at a traffic incident scene. This is especially true for major traffic incidents that might involve a number of emergency vehicles. If good traffic control is established through placement of advanced warning signs and traffic control devices to divert or detour traffic, then public safety agencies can perform their tasks on scene with minimal emergency-vehicle lighting.

Guidance:
Public safety agencies should examine their policies on the use of emergency-vehicle lighting, especially after a traffic incident scene is secured, with the aim of reducing the use of this lighting as much as possible while not endangering those at the scene. Special consideration should be given to reducing or extinguishing forward facing emergency-vehicle lighting, especially on divided roadways, to reduce distractions to oncoming road users.

The MUTCD also addresses cone placement as follows:

Section 6F.56 Cones
Standard:
Cones shall be predominantly orange and shall be made of a material that can be struck without causing damage to the impacting vehicle. For daytime and low-speed roadways, cones shall be not less than 450 mm (18 in) in height. When cones are used on freeways and other high-speed highways or at night on all highways, or when more conspicuous guidance is needed, cones shall be a minimum of 700 mm (28 in) in height. For nighttime use, cones shall be retroreflectorized or equipped with lighting devices for maximum visibility. Retroreflectorization of 700 mm (28 in) or larger cones shall be provided by a white band 150 mm (6 in) wide located 75 to 100 mm (3 to 4 in) from the top of the cone and an additional 100 mm (4 in) wide white band approximately 50 mm (2 in) below the 150 mm (6 in) band.

Option:
Traffic cones may be used to channelize road users, divide opposing motor vehicle traffic lanes, divide lanes when two or more lanes are kept open in the same direction, and delineate short duration maintenance and utility work.

REFLECTIVE VESTS

Personnel visibility is critical during highway operations, especially at night. Being struck by a vehicle is a constant danger. Dawn, dusk, and inclement weather compromises visibility and increases the risk of being struck. It is imperative that workers be visible as a person among the flashing lights and other apparatus marking. ANSI/ISEA Standard 107-1999 specifies the minimum amount of fabric and reflective materials to be placed onto safety garments that are worn by workers near vehicular traffic. This standard is now the most commonly used standard associated with safety vests.

Reflective vests should be used to increase worker visibility; regardless of the use of turnout gear. The reflective vests used by emergency workers should have both retroreflective and florescent properties. Retroreflective material returns the majority of light from the light source back to the observer. Florescent material absorbs UV light of a certain wavelength and regenerates it into visual energy. Plano Fire Department requires all personnel involved in highway operations to wear a Class II vest with a lime green background and orange/gray stripes, shown in Figures 5-14 a and b.

Figure 5-14 a and b. A Plano Fire Department traffic vest.

STANDARD OPERATING PROCEDURES

Seven fire departments in Northern Virginia, have formed a regional committee to review highway safety policies and recommend changes to them when necessary. The committee is made up of the following departments:

- Fairfax County Fire and Rescue Department;
- Arlington County Fire Department;
- Fort Belvoir Fire Department;
- Fairfax City Fire Department;
- Reagan National Airport Fire Department;
- Dulles Airport Fire Department; and
- Alexandria County Fire Department.

Any member of any of the participating fire departments can submit a proposed change. The proposed change is passed through the chain of command to the committee for review. In Fairfax County, the Safety Officer and an operations chief review the proposed change before it is advanced to the committee. Committee recommendations go to all department chiefs, who ultimately decide if the recommendation will be implemented in the individual department. The regional committee does not mandate changes to individual departments. It serves primarily as an information sharing forum to improve interoperability among participating departments.

The Virginia Beach Fire and Rescue SOP, *Operations of Fire Department Vehicles*, was in the process of revision at the time of this writing. Any member of the department can submit suggested policy changes. Suggestions go first to a specific committee, such as the Safety Committee or Apparatus Committee, for review. If the suggestion has merit, it is forwarded to the SOP committee. The SOP committee forwards its final recommendation to the chief for a decision on implementation. It is not unusual for changes to go back and forth between the chief and the SOP committee several times until consensus is reached. After approval, the SOP is implemented as an interim policy for 90 days for evaluation and any necessary modification before it becomes final. The Fairfax County Fire and Rescue *Operating Procedures for Highway Incidents* and the Phoenix Fire Department *Safe Parking While Operating in or Near Vehicle Traffic* can be found in Appendix C.

Effect on Highway Operations

The design and insulation of new passenger and commercial vehicles, combined with operating radio/CD and air conditioning, can significantly hinder a driver's ability to hear emergency vehicle audible warning devices and realize emergency vehicles are approaching. Departments must engrain safety into all activities; and personnel must view safety as their job, both individually and collectively.

No department interviewed had any specific data to support current apparatus placement or lighting patterns. However, personnel had several subjective comments regarding the effects. All Phoenix Fire Department field personnel, including chiefs, believed that crew communication and apparatus positioning were the primary methods of providing a safe operational environment.

Plano Fire Department personnel also unanimously agreed that apparatus placement is the most effective of all methods used on scene to improve safety of providers. They believed the department's placement policy definitely improved safety and decreased near-misses. Blocking a lane with apparatus slows traffic speed and impedes traffic flow. The blocking policy increases the feeling of safety for both fire and law enforcement personnel, but all remain vigilant. Even when proper blocking is used, personnel have witnessed drivers going over curbs, cones, and even between blocking apparatus. Vehicles have struck blocking apparatus. Had the apparatus not been there, the vehicle could have breached the work zone and injured personnel.

All the departments placed less importance on lighting patterns versus apparatus placement. Phoenix fire personnel considered lighting secondary and, in many cases, unreliable. Some believed that any lighting pattern was useless at night when extensive floodlights were used. Some Fairfax County personnel believed that directional signaling (light sticks, arrows, etc.) seemed to help traffic flow, but identified pulling all the information together and discipline as the keys to successful highway operations, and rapid mitigation as the key to improving and restoring traffic flow.

Limitations

Plano Fire Department personnel identified forgetting to shut off apparatus headlights and civilian motorists paying more attention to what is going on at the scene than to traffic flow as their biggest problems during highway operations. Another challenge is dealing with neighboring districts that do not use similar apparatus placement practices, which can result in confusion and policy issues when responding with mutual aid. In an effort to reduce this problem, Plano shared its policies with neighboring departments. It must be remembered that SOP's and operating manuals serve as a basis for safe practice. Actual practices depend on the existing conditions identified by the company officer or Incident Commander.

RECOMMENDATIONS FOR HIGHWAY OPERATIONS

- During highway operations, position the engine at a 45-degree angle to the lanes, with the pump panel toward the incident and the front wheels rotated away from the incident.
- Extinguish forward-facing emergency vehicle lighting, especially on divided roadways.
- Reduce the use of lighting as much as possible at the scene.
- Require members to wear highly reflective material when conducting highway operations.
- Remain vigilant during all phases of highway operations.
- Work with neighboring districts to develop similar highway operations policies.
- Allow all members to submit suggested policy changes.

---

[1] National Bureau of Standards. (1978). *Emergency Vehicle Warning Lights: State of the Art*, Pub. No.480-16.

[2] Solomon, S. (2002). The case for amber emergency warning lights. *Firehouse* (2), 100-102.

[3] Ibid.

RESPONSE

INTRODUCTION

Responses that are true emergencies (both fire and EMS) are limited. Yet, tradition dictates that apparatus respond using lights and siren to all calls received through 911. Responding in the emergency mode increases the risk for vehicular crashes resulting in injuries and fatalities to both emergency responders and civilians. This section reviews practices that decrease that risk by limiting emergency response (use of lights and siren).

PRIORITY DISPATCH

The use of warning lights and sirens on fire and EMS emergency vehicles is a basic component of emergency response and patient transport in this country. Over the past several years, the effectiveness of this long-standing tradition in affecting patient outcome or decreasing property or financial loss has come into question. What is known is that the majority of emergency vehicle crashes occur when warning lights and siren are in use. Multiple studies have been conducted comparing ambulance response times with and without the use of lights and sirens. Separate studies conducted by Addario, et al. and Kupas, Dula, and Pino have demonstrated that, although response times are faster with lights and siren, the time saved had no significant impact on patient outcome, except in cardiac arrest and obstructed airway.[1,2] Because of the risk associated with lights and siren, the National Association of EMS Physicians (NAEMSP) and the National Association of State EMS Directors (NAEMSD) published the position paper, *Use of Warning Lights and Siren in Emergency Medical Vehicle Response and Patient Transport*, in 1993. This document is available online at: http://www.naemsp.org/Position%20Papers/WarnLghtSirn.html

If a dispatcher gets accurate and complete information from the caller, he/she should be able to determine if the condition warrants a lights and siren response from the information given. Sound priority dispatching relies on trained dispatchers and preset, algorithmic questions. There are commercial programs with preset questions available for both fire and EMS conditions. Departments also have developed their own preset questions.

Responses within Salt Lake County, with the exception of Salt Lake City, are automatic and cross jurisdictional boundaries. All calls are dispatched based on the closest station rather than the jurisdiction of the incident location. Testing of Automatic Vehicle Locator (AVL) dispatching began in May 2003. All fire apparatus have Global Positioning Satellite (GPS) locators installed. During the testing phase, units are dispatched using both the closest unit (AVL) and the closest station. Data are being collected to compare and assure the closest unit is actually being dispatched. Response times will also be compared between the two systems. A plan is in place to dispatch a second unit to assure the first AVL unit dispatched does not remain out of its jurisdiction on consecutive calls.

Priority dispatch for both fire and EMS has been part of the Salt Lake County system for all departments for 25 years. Priority dispatch began with Dr. Jeff Clawson's Emergency Medical Dispatch (EMD), which he developed while Medical Director for Salt Lake City Fire Department. Over time, the concept was extended to fire calls. Currently Salt Lake City uses Clawson's system, which must be purchased. The nine departments that provide services for the rest of Salt Lake County organized a committee of representatives, including chiefs, who developed the priority questions that are used by VECC to determine priority dispatch codes. An example of a Call Guide used by dispatch can be found in Figure 6-1, below.

SAMPLE VALLEY EMERGENCY COMMUNICATIONS CENTER CALL GUIDE

HAZFIRE: Hazardous material spill or leak involving a fire.   Priority 1

DEFINITION: Any hazardous material spill or leak involving a fire.

CALL-TAKER PROCEDURES:

1. Line #.
2. Location of occurrence.
    1. Name or type of building or structure.
    2. Apt #, building #, suite #, as applies.
    3. Specific location of the problem and access information for responding units.
    4. Is the fire inside or outside?
    5. Is this a drug lab?
3. Is it the building itself or a hazardous material that is on fire?
    If this is a hazardous material that is on fire, what is the material?
4. Are you inside the building where the fire is occurring?

YES	NO/ or Complainant Callback
• If safe to do so, follow your evacuation protocol and get everyone out. • Take the safest way out and then DO NOT go back in. • Where is your safe location or meeting place? • If possible, have someone call back 911 on a cell phone.	1. Are you at a safe location? If yes, go to question #5. If no, Go to a safe location and call us back on 911.

5. Do you think anyone is still inside? (Cars in the parking lot, lights on?)
6. Are you still near the building?
7. Where is everyone meeting?
8. Does anyone need medical help?
    1. If yes, refer to medical priority cards.
    2. Advise the fire dispatchers.
9. How many people were injured or affected?
10. If complainant does not know what the material is, ask;
    1. Is there a placard on the container?
       If yes, get the color of the placard and the number on the card. If this is a railroad tanker and it is safe to do so, get the number off of the side of the tanker.
    2. If the substance is not known, is it a liquid, solid, or vapor?
    3. Where did it come from? (Source: bucket, barrel, truck, or tanker?)
    4. How much was spilled?
    5. Is the wind affecting the situation?
       1. If so, what direction is the wind blowing?
    6. Is the substance threatening or leaking into any waterways, storm drains, or sewers?
11. Complainant information (include name, address, and phone of where they are calling from).

Figure 6-1. Sample Valley Emergency Communications Center Call Guide.

SAMPLE VALLEY EMERGENCY COMMUNICATIONS CENTER CALL GUIDE (CONTINUED)

DISPATCHER PROCEDURES:
1. Refer to the individual agency's fire assignments for the appropriate dispatch response.
2. Anywhere other than Sandy or West Valley, fire dispatchers MUST notify the County EOC Duty Officer (LEPC) in a timely manner, if this is:
    1. Confirmed spill or leak of over 50 gallons of gasoline or hydrocarbon.
    2. Any spilled or released hazardous material threatening or in a waterway.
    3. Any spilled or released hazardous material from a business or facility.
    4. Abandoned, discarded, or unknown hazardous materials.
    5. Spills or leaks occurring during transport.
3. If thi ccurs in Sandy or West Valley then you MUST notify Sandy and West Valley's LEPC for all of the above types of incidents.
4. If, upon arrival, units advise this is a working fire:
    1. Assign WF to the primary unit.
    2. Dispatch any other required equipment, when applicable.
    3. Notify Utilities.
    4. Make necessary notifications to Command Personnel.
    5. Dispatch any other required equipment when applicable.
    6. Continue time notifications until told otherwise.
5. If it is discovered that there are injuries before units arrive on scene, advise the primary unit and ascertain if any additional units should be sent. **LEPC = Local Emergency Planning Committee. Notifying the LEPC complies with the Federal law, SARA III.

Updated: 09/28/99

Figure 6-1. Sample Valley Emergency Communications Center Call Guide (continued).

In 1997, the Virginia Beach Fire and Rescue Safety Officer, City Risk Manager, and Deputy Chief of Operations identified a potential problem while prioritizing department risks using the risk matrix in Figure 6-2. Vehicle and intersection crashes were reviewed and compared to the type of call dispatched. It was found that most of the incidents were low frequency, low severity. The safety officer also reviewed the St. Louis, Missouri, Fire Department's "On-The-Quiet" policy at the time. This group used the findings to classify certain incident types as nonemergency responses.

Using the classifications, the Safety Officer worked with the SOP Committee to develop the priority dispatch policy, which was approved by the risk manager, the deputy chief, the chief, and finally, the city attorney. Table 6-1 (on page 44) contains the classifications of emergency and nonemergency fire responses.

Phoenix fire apparatus are initially dispatched to respond with lights and siren on the majority of calls. If the first-arriving unit finds nothing on approach, the remaining responding units are downgraded to continue without lights and siren. Virginia Beach Fire and Rescue uses the same approach on multiunit emergency responses. If an emergency situation does not exist, the first-arriving unit transmits a radio downgrade to all responding units. This downgrade transmission includes a brief description of findings. All other responding units continue to the scene in non-emergency mode, until cleared by command.

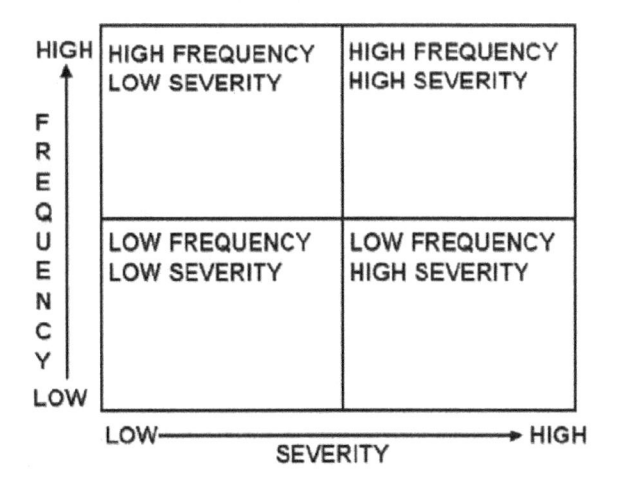

Figure 6-2. Risk Assessment Matrix.

TABLE 6-1.
VIRGINIA BEACH FIRE DISPATCH RESPONSE PRIORITIES

Emergency	Nonemergency
Structures, commercial or residential	Lockin or lockout (weather conditions and patient location may require upgraded response)
Vehicle fires on the interstate	Public service call and/or assisting police
Brush	Unauthorized burning
Boat(s)	Controlled burning
Rubbish or dumpster	Vicinity alarms
Aircraft	Citizen's complaint
Commercial and residential fire alarm(s)	Fuel spills where the product is not an immediate threat to life or environment
Electrical wires down	Carbon monoxide detector activation (without signs of CO poisoning)
Residential/Commercial flammables	Water leaks
	Move-ups
	Manpower assistance
	Salvage truck, unless upgraded by command
	Bomb threats
	Wires hanging -- not on fire
	Any incident downgraded by the on-scene resource

Certain conditions are dispatched nonemergency (no lights and siren.) Appendix D contains all the nature codes used by the Phoenix Fire Department. Conditions where no default code is indicated require the call taker to specify Code 2 (nonemergency) or Code 3 (emergency lights and siren). All company officers have the authority to change the level of response and are encouraged to err on the side of caution and firefighter safety. If the code is changed, it is recorded in the dispatch center records by the Mobile Data Terminal (MDT) transmission. None of the departments visited had any data on the number/percent of responses where the response code was changed by the company officer.

As all issues are within the Phoenix Fire Department, a labor-management committee, with input from the medical director, determined the nonemergency response EMS conditions. The EMS dispatch codes were developed about 1998 and, as of this writing, had not undergone a review. Units dispatched in the nonemergency mode can be preempted and redirected to an emergency response.

The average response time for a Phoenix nonemergency dispatch is 7 minutes, and just under 5 minutes for an emergency dispatch. The dispatch supervisor believed that most increases in response times are a result of annexation and larger coverage areas. If a response time exceeds 10 minutes, dispatch receives an electronic notice and follows up by contacting the unit to determine the cause.

RESPONSE MATRIX

With the exception of Salt Lake City, all the fire departments in the Salt Lake Valley belong to a user alliance. The alliance meets on a monthly basis. They developed a recommended response matrix for each incident type, which was sent to each department's chief for approval. Each department chief had the option to disapprove or modify the response matrix. The response matrix for each city is maintained in the VECC. An example from the matrix is found in Figure 6-3. The matrix is updated and modified based on dispatch feedback forms and changes in technology (Figure 6-4). The entire fire response matrix and the mass casualty response matrix for the Salt Lake valley departments can be found in Appendix E.

The Virginia Beach department's response matrix was revised April 1, 2003. The policy now includes two engines responding from opposing directions to all accidents on the interstate. Members cannot jump or navigate jersey barriers dividing the highway in order to reach an accident scene from the opposite lane. Apparatus proceed in a manner that allows them to position in the lane and direction of travel where the incident occurred.

EXAMPLE OF SALT LAKE VALLEY FIRE RESPONSE MATRIX									
CODE	BDALE	COUNTY	MIDVALE	MURRAY	SANDY	S JORDAN	SOSL	W JORDAN	W VALLEY
ALARM	1E	1E	1E	1E	1E	1E,1A	1E	1E	1E
ALRMCO	1E	1E	1E	1E	1E	1E	1E	1E	1E
ALRMR	1E	1E	1E, CS	1E	1E	1E, 1A	1E	1E	1E
ALRMW	1E, 1T	1E, 1T	1E, 1T	1E, 1T	1E, 1T	1E, 1T, 1A	1E, 1T	1E, 1T	1E, 1T

Figure 6-3. Example of Salt Lake Valley fire response matrix.

DISPATCH FEEDBACK FORM

ADMINISTRATIONS: 840-4100 / FAX: 840-4040   DISPATCH: 840-4000 / FAX: 840-4039

Date Received: _____ VECC CAD#: _____ Date Assigned: _____

Date Completed: _____ VECC Case#: _____ VECC Supervisors Initials: _____

SALT LAKE VALLEY EMERGENCY COMMUNICATIONS CENTER DISPATCH FEEDBACK REPORT

DATE: _____ REQUESTING AGENCY: _____

REPORTED BY: _____ PHONE #: _____

INCIDENT DATE: _____ TIME: _____ CARE #: _____

INCIDENT TYPE: _____ LOCATION: _____

PROBLEM/CONCERN: _____

_____

_____

_____

_____

Would you like the response: In writing?_____   Verbally?_____

Figure 6-4. Dispatch feedback form.

Responding engines that are not on the incident side of the highway contact the officer of the apparatus that is to find out if the unit will be needed, should stage until further evaluation is completed, or should clear. If the unit is needed, it continues to the next exit and reenters the highway system to assist from the same lane and direction. If the unit is asked to stage, it proceeds to the next exit and stages at an appropriate location until cleared. The Virginia Beach Fire and Rescue fire response matrix also can be found in Appendix E.

VOLUNTEER RESPONSE IN PRIVATELY OWNED VEHICLES (POV's)

It is common practice for members of volunteer fire and EMS departments to respond to the station (and, in some instances, to an incident location) using their private vehicles. Many States allow volunteer members to equip their POV with emergency lights. Acceptable color for those lights varies from State to State. Some States allow only red lights, some allow only blue lights, and some allow both. In addition, departments may specify the color based on the rank held by the individual (e.g., red is used only by chief officers.) Regardless of color, a review of the data in the USFA *Firefighter Fatalities in the United States* reports from 1990 to 2000 showed that 25 percent of firefighters who died in motor vehicle crashes were killed in privately owned vehicles.

Although volunteers can legally run with lights on a POV, the State of Utah requires them to have authorization from local law enforcement and a rider on their personal insurance policy. Therefore, volunteers in the Salt Lake Valley do not run with lights on POV's. There have been no issues related to response times because they travel with normal traffic when responding to the station.

RECOMMENDATIONS FOR RESPONSE
- Assess and code responses using the classic risk management matrix.
- Respond to low frequency, low severity incidents in a nonemergency mode.
- Train dispatchers in the use of preset questions to determine emergency versus nonemergency response.

---

[1] Addario, M., Brown, L., Hogue, T., Hunt R. C., & Whitney, C. L. (2000). Do warning lights and sirens reduce ambulance response times? *Prehospital Emergency Care, 4* (1), 70-4.

[2] Kupas, D., Dula, D., & Pino, B. (1994). Patient outcome using medical protocol to limit "lights and siren" transport. *Prehospital and Disaster Medicine. 9* (4).

TRAINING

INTRODUCTION

Through attitude and behavior, organization leaders must reflect the importance of safety in all aspects dealing with vehicles. This attitude must be infused in all organizational policies and training. Department commitment to driver competency and accountability can have a profound effect on reducing crashes, injuries, and fatalities. This begins with the factors used to select the driver candidate.

Training is the foundation of all safe practices. The type of course, integration of classroom and applied practice, and instructor qualifications all contribute to the effectiveness of any training. This section reviews factors to consider in selecting driver candidates, practices, and programs related to several aspects of training (including initial basic driving courses); use and value of a driving simulator in basic, remedial, and continuing training; and developing driver requirements.

HUMAN FACTORS

The participants in the forums for the EVSI identified four categories of human factors that contribute to vehicle crashes: knowledge base, skills, ability, and attitude. Drivers may lack knowledge of traffic laws, physical laws that govern the apparatus operation, or may lack awareness of potential dangers. Inadequate skills in handling apparatus may be the result of insufficient training, lack of hands-on training, inexperience, slowed or improper reaction, or poor judgment. Attitude plays a major role in safe vehicle operations. Failure to obey laws or take proper precautions, improper use of the roads, allowing excitement to lead to impulsive actions, dangerous shortcuts, and irresponsible or reckless behavior all contribute to apparatus crashes and fatalities. Other factors include inattentiveness, failure to concentrate on driving tasks, and the emotional sense of power and urgency when running lights and siren. This sense can block out reason and prudence, leading to the reckless operation of the emergency vehicle.

A review of the USFA *Firefighter Fatalities in the United States* reports suggests that age (both young and older) appears to be a factor in both fire apparatus and ambulance crashes. Many who enter the emergency services field have only 2 to 3 years of normal driving experience. The older driver may have slowed responses. The forum participants suggested setting a minimum age of 21 for drivers. In some cases, the organization's insurance company may determine this.

Another factor that contributes to emergency vehicle crashes is improper or no background checks of the potential drivers. According to an article in the Detroit News, 41 percent of ambulance drivers involved in fatal accidents had prior citations on their driving records.[1]

Prior to employment, the department should verify driving records by having the applicant sign a consent form to allow the department to check records with the State. The applicant's driving record should be reviewed for the number of moving violations over the past 36 months, any driving under the influence (DUI) convictions, reckless driving citations, license suspensions, etc. Evaluation of driver education certificates, specialized driving license (e.g., CDL), and physical qualifications (e.g. vision and hearing) can also help reduce the number of vehicular incidents that occur each year. The American Ambulance Association (AAA) has developed a Best Practice for EMS Driving Outline, which can be found in Appendix F.

DRIVER TRAINING COURSES

During the course of this study, several fire apparatus driver training programs were reviewed. Information on these programs is discussed in the following section.

California Department of Forestry and Fire Protection

The CDFFP Academy's Basic Fire Engine Operator (BFEO) driver training program began in the 1980's as more emphasis was placed on emergency response. The BFEO driver program was replaced in 2000 with the current driving course. This change occurred as a result of several serious, avoidable apparatus crashes. Although there were no fatalities, there were some serious injuries. In all cases, the incidents were found to be driver error. Near-misses also played a part in the development of the driving program. CDFFP investigators found that approximately 90 percent of vehicle crashes occurred at intersections. The current program was customized for CDFFP's needs using components of the Peace Officer Standard Training (POST) and other successful driving programs.

Full time CDFFP drivers attend 10 weeks of training at the Academy. This consists of:

- Orientation   1 week;
- Structures and Rescue   3 weeks;
- Driver operations (includes pump operations)   3 weeks; and
- Wildland (includes apparatus placement on the line)   3 weeks.

CDFFP also uses a large number of seasonal workers during the 6-month fire season. If a seasonal worker shows above-average performance, the department may bring him/her in to complete just the driver operations module, which leads to classification as a Limited Term Engineer and allows the seasonal worker to drive the apparatus.

The average driver operations course consists of 30 participants with an instructor-to-student ratio of 1:4 to 6. This course is conducted entirely by the three staff instructors and two mechanics. The driver operations course consists of 132 hours, broken down as shown in Table 7.1.

In addition to the driver operations course, the CDFFP Academy teaches the Heavy Fire Equipment Operator (HFEO) course. This is a 6-week course that covers the operation of dozers, truck/trailers, etc. It includes the same EVOC component as the driver operations course. It is taught by both staff and outside heavy equipment subject matter experts.

Drivers felt the operations course was more intense than the old BFEO program, placed more emphasis on theory, and had more active instructor involve-

TABLE 7.1 CDFFP DRIVER OPERATIONS COURSE BREAKDOWN	
ORIENTATION	
Orientation	3
Written Examinations	5
Performance Exams	8
Clean-up	12
Total	28
PHYSICAL FITNESS	
Physical Training Sessions	8
Total	8
PUMP OPERATIONS	
Pump Theory	6
Hydraulics	2
Pump Skills	17
Total	25
VEHICLE OPERATIONS	
Basic Driving and Air Brakes	6
Preventive Maintenance	13
Emergency Vehicle Operations	14
Checkout Drives	8
Cross Country Driving (Field Exercise)	8
Off Road Vehicle Operations	10
Total	59
EMERGENCY/FIREGROUND OPERATIONS	
Introduction to I-Zone	4
Multi-Co Drills	8
Total	12
TOTAL	132

ment. They agreed that including more theory and discussion of vehicle dynamics improved their knowledge of apparatus operation and their decisionmaking ability by helping them learn to anticipate instead of just react. They cited the dogleg course as one example and commended the addition of the Interface Zone (I-Zone) training. An I-Zone Exercise Evaluation can be found in Figure 7-1. The objectives and specific hours breakdown for each topic in the Vehicle Operations course can be found in Appendix G.

I-ZONE EXERCISE EVALUATION

STUDENT _____ ENGINE NO _____

SCENARIO NO _____ DATE _____ TIME _____

CREW OPERATIONS		POINTS	POINTS
1	Provides safety briefing to crew	2	
2	Gives clear instructions/assignments to crew	2	
3	Retreat signal identified with crew	2	
4	Hoselines deployed appropriately for scenario	2	
ENGINE OPERATIONS			
1	Doors and windows are closed	2	
2	Code 3 and driving lights are on	2	
3	Engine is backed in	2	
4	Engine does not block access to other vehicles	2	
5	Engine protection line in place and charged	2	
6	Garden hose placed in tank	2	
7	Ladder and roof protection lines placed in service	2	
8	Not parked next to wood pile or debris	2	
9	Engine positioned so propane tank not a factor	2	
10	Not parked under powerline	2	
STRUCTURE			
1	Windows and doors closed	2	
2	Propane tank shut off	2	
3	Hazardous vegetation identified and discussed	2	
4	Hazardous wood pile or debris identified and discussed	2	
5	Out buildings are determined and contents indentified	2	
PUBLIC CONSIDERATIONS			
1	Number of occupants or residents are determined and contact made. Animals are identified and planned for	2	
	TOTAL	40	

PROCTOR _____
(Please use back for comments if necessary) Comments on back ☐
00012801 evl.doc
February 26, 2003
Page 1

Figure 7.1. CDFFP I-Zone evaluation.

STUDENT REQUIREMENTS

The driver operations course is open to driver trainees only. Driver trainees must be at least 18 years old and hold a Class B license. Since most trainees spend 6 to 7 years as firefighters first, the mean age of CDFFP drivers is mid-20's.

The training chief identified a change in member demographics from an agricultural background with experience in driving tractors and bigger equipment to a college background. This shift, along with the fact that many of the trainees do not obtain the Class B license until a month or so before the class starts, contributes to a growing pool of students having little to no experience with bigger equipment.

INSTRUCTOR QUALIFICATIONS

All Academy instructors must be California Instructor 1A and 1B trained. The instructor training program is based on the International Fire Service Training Association (IFSTA), *Fire and Emergency Services Instructor* manual, which in turn is based on NFPA 1041, *Standard for Fire Service Instructor Professional Qualifications*. Each component is 40 hours. California Instructor 1A covers manipulative lesson plans; 1B covers technical lesson plans. This training is offered by the State Fire Marshal's office. All Academy instructors are also DL170 agents of the Department of Motor Vehicles (DMV) Employer Licensing Unit. The DMV audits the DL170 records for CDFFP. The testing course must be mapped out and approved by the DMV annually.

There is no standard driver training on the unit level, and these requirements do not apply to the unit driving instructors. Some units do yearly EVOC driving days, but others do not. Most of the unit training officers have received instructor training, but State certification is not required.

RECERTIFICATION /REMEDIAL TRAINING

CDFFP currently has no requirement for recertification or repeating any portion of the driver operations course. The drivers interviewed believed that the course should be used to recertify, if it included more advanced scenarios. Although remedial training is normally handled at the unit level, unit officers can send drivers back for remedial training if they believe the driver's skills indicate this intervention.

Sacramento Regional Training Facility

In 1993, Sacramento City Fire implemented ALS. Paramedics were recruited and trained at the fire academy, but the training did not include emergency driving. The accident frequency rate was 90 at-fault ambulance motor vehicle crashes per one million miles the first year of the ALS program. This resulted in a significant increase in third-party insurance costs, in particular on the medic units.

In 1994, the Risk Management, Loss Control Safety Unit coordinated an interdepartmental EVOC between police and fire to train and certify fire EVOC instructors. After EVOC became mandatory for completion of the fire academy in 1995, a 39-percent reduction in ambulance at-fault crashes occurred.

The Sacramento Regional Training Facility provides driver training for all levels of personnel: recruits, firefighters, drivers, captains, support staff, law enforcement officers, etc. The facility has an instructor-to-student ratio of 1:2. The facility develops training programs by evaluating multiple programs and customizing courses to meet the needs of the Joint Powers partners. For example, the fire instructors built a customized off-road course by adding components to the existing CDFFP and Forest Service courses.

The primary course consists of 8 hours of classroom for defensive driving, 16 hours of actual course driving, and 4 hours in the simulator. The driving portion concentrates on vehicle placement and the physics of dynamics. An example of a lesson plan used for smaller vehicles can be found in Appendix G.

Fire personnel train only during daytime hours, while police train during both day and nighttime hours. Recruits spend approximately 90 percent of their 40-hour program behind the wheel. Vehicles used for training include police sedans, SUV's, engines, ambulances, a bus, off-road equipment, and trailer trucks. Personnel drive both their assigned apparatus and training apparatus.

INSTRUCTOR QUALIFICATIONS
All instructors are Fire Course Instructors IA and IB and POST certified, have completed the 40-hour National Safety Council course, and an 8-hour class on law and liability issues. Although the goal is 100 percent, currently 75 percent of the instructors are certified to conduct DMV training and administer DMV examinations according to California law DL170. All instructors are cross-trained, which allows fire instructors to train police personnel and vice versa.

RECERTIFICATION/REMEDIAL TRAINING
Fire personnel repeat the entire primary training course each year as a mandatory recertification. In addition, the Accident Review Committee can refer anyone found responsible in an at-fault incident to remedial training. Supervisors can also request remedial training for personnel. Remedial training is not a punitive action. It is used to reinforce training and is done in a positive manner.

Ventura County (California) Fire Department
Ventura County Fire Training conducts seven driving courses:
- Class A Driver Training;
- Paramedic Squad Driver Training;
- Class B Driver Training;
- Off-Road Driver Training;
- Class C Driver Training;
- Driving Simulator; and
- Tillered Truck Driver Training.

A description of each of these courses and the instructor qualifications can be found in Appendix G.

INSTRUCTOR QUALIFICATIONS
In addition to the instructor qualifications listed for each course, all instructors are California Fire Instructor IA and IB trained, POST certified, and certified to conduct DMV training and administer DMV examinations under the DL170 program. They completed the Los Angeles County police simulator training and received safety officer training.

DRIVING SIMULATOR
Both Sacramento Regional Training Facility and Ventura County Fire Department have driving simulators. At the time of the site visit, Sacramento had been using the simulator for 4 years. Ventura County had the simulator for 2 years, but the first year was fraught with technical problems, so it had only been used for 1 year.

The Sacramento Regional Training Facility uses the driving simulator strictly to improve drivers' decisionmaking processes. A simulator cannot teach the basic driving skills acquired in training exercises using actual emergency vehicles. Time has shown that errors in decisionmaking are primarily related to intersection clearing and conflict resolution (obtaining the right of way). The decisionmaking process is reinforced as personnel complete 4 hours in the simulator as part of the annual recertification process.

One of Ventura County's goals for the simulator program is to educate drivers in safe driving techniques. Initially, the emphasis was on emergency driving and backing scenarios. In March 2003, they began using the scripting tool to develop scenarios that address intersection analysis, conflict management, and desired apparatus placement options at an incident. Because the simulator was new, the Training Section was working to overcome the "video game" mentality that line personnel were exhibiting and provide the rationale to gain buy-in.

Technical Aspects
Ventura County uses a three-station mobile simulator manufactured by FAAC, Inc. At the time the unit was purchased, the simulator cost approximately $500,000, plus $20,000 for the tractor. It is housed in a 42-foot 5th

wheel RV trailer so it can be moved from location to location. Ventura County also holds a service/maintenance agreement on the simulator, which cost $10,000 annually (Figure 7-2). They believed the additional cost to make the simulator mobile (approximately $100,000) provided a cost savings compared to stationary training by eliminating the following:

- fuel for units to travel to training;
- wear and tear on apparatus;
- time saving; and
- cost of a cover assignment from another company.

The simulator contains three stations that can be configured to represent fire engines, EMS, or other types of vehicles (Figure 7-3).

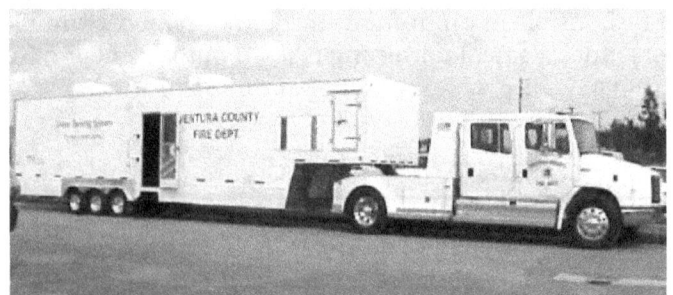

Figure 7-2. Ventura County Fire Department mobile simulator.

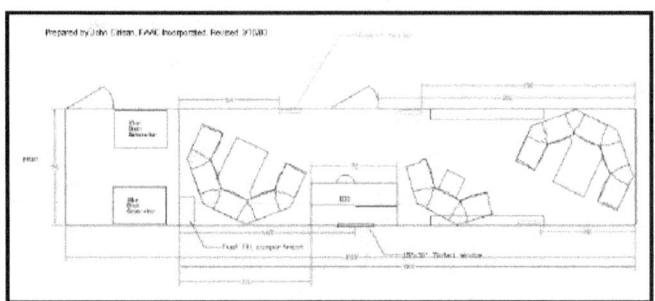

Figure 7-3. Internal layout of three stations.

This particular simulator has five screens in each station. Multiple stations are linked together through the instructor console for interactive scenarios. Each station has headsets that allow the captain and driver to communicate. The headsets can be programmed with radio and background noise to simulate real conditions. The stations are configured to interchange vehicle cab models (e.g., sedan, engine, truck, etc.) This mobile simulator has two engine cab model stations and one sedan cab model station (Figure 7-4). The manufacturer redesigned the engine cab models to match the dash of the Ventura County engines. Regardless of the cab model, the physical response of the steering wheel, brakes, etc., for any desired vehicle can be programmed into the computer feedback.

While the physics affecting all vehicles is the same, the specific apparatus response to forces (e.g., acceleration speed, braking resistance, distance, speed, evasive maneuvers, etc.) is programmed in as needed for a given scenario. Vehicle sounds of acceleration, deceleration, tire squeal, siren, radio communication with other simulator "vehicles," weather, and day/night conditions also can be programmed from the instructors' control console (Figure 7-5).

Figure 7-4. Engine cab model station.

Figure 7-5. Instructor control console.

This simulator has four virtual worlds of operation, including urban and rural areas (Figure 7-6). The programs allow the instructors and students to view the apparatus position from an overhead vantage point (Figure 7-7). This feature is useful during the scenario replay to show apparatus placement in relation to curbs, etc. The manufacturer also programmed the graphics by incorporating pictures of actual Ventura County apparatus, so that is what is seen in the overhead view.

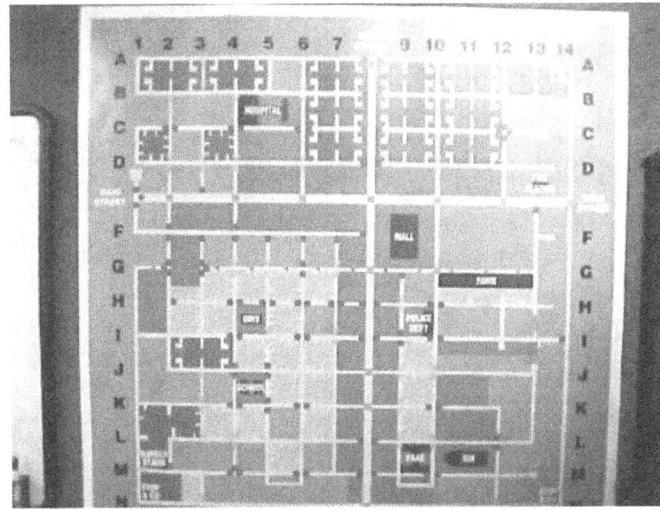

Figure 7-6. Virtual world map.

Figure 7-7. Overhead view

The Sacramento facility uses a four-station fixed Advanced Mobile Operations Simulator (AMOS) manufactured by Doron that cost approximately $750,000 at the time of purchase. This particular simulator has three flat plasma screens in each station (Figure 7-8). Each screen costs approximately $5,000. Just as the mobile simulator at Ventura County, the stationary simulator allows linking of multiple stations for interactive scenarios, programming vehicle physical responses, sounds, radio communication with other simulator "vehicles," and altering weather and day or nighttime driving through the instructor console. The stations are also equipped to interchange vehicle cab models (e.g., sedan, engine, truck, etc.). However, because vehicle dynamics are the same for a vehicle whether small or large, and only the end result differs, the only cab model used by the Regional Training Center is that of a sedan. Additional cab models are priced

Figure 7-8. Three-screen simulator station with sedan cab.

separately from the unit. A motion base can be added to cab models to provide realism in road feel for an additional fee.

Training Sessions

A relatively common side effect of "driving" the simulator is Simulator Adaptation Syndrome (SAS). SAS creates motion sickness symptoms. Ventura County experienced an incidence of 5 to 7 percent of members unable to complete the simulator exercises due to SAS. Sacramento Regional Training Facility experienced an incidence of approximately 40 percent of participants developing some motion sickness symptoms.

There are a variety of methods used to reduce the incidence of SAS, with varying degrees of success. Ventura County requires all members to wear special wristbands while driving the simulator. Sacramento uses both the special wristbands and special glasses (Figures 7-9 a and b). Limiting the time the member is actually in the simulator is the most effective way to control SAS.

Each simulator session lasts 3 to 4 hours and involves a total of 20 to 30 minutes of actual simulator driving time. The session begins with an orientation process to learn to "drive" the simulator (acclimate to the cab model and graphics). The acclimation portion puts the driver through four different scenarios. Each scenario lasts approximately 2 minutes. The driver is removed from the simulator for 5 to 6 minutes following each completed scenario.

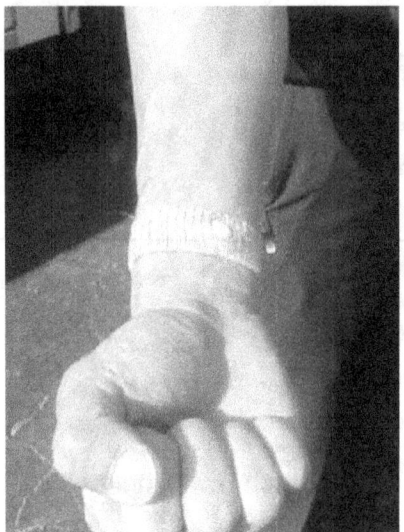
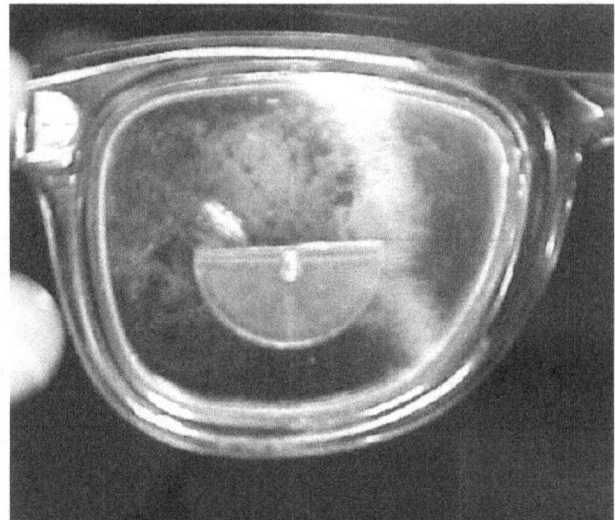

Figures 7-9 a and b. Methods of controlling SAS.

Ventura County was beginning to build their scenario bank at the time of the site visit. The remaining scenarios used at the Sacramento facility reflect situations requiring rapid decisionmaking. The facility has a bank of approximately 400 usable scenarios. Most classes use 18 to 25 scenarios that focus on the most common situations with high-risk potential. These situations are usually intersection clearing and conflict resolution. If the driver makes an incorrect decision, the feedback on the screen is immediate (Figure 7-10). Both simulators allow the instructor to play back a driver's performance through the entire scenario for review and teaching purposes.

Currently, the simulator is not used as part of a recertification program or for any type of remedial training by Ventura County. Training's short-term goal is to send all personnel through a 3-hour simulator session biannually to reinforce good driving judgment. A long-term goal is to implement a nonpunitive annual driver recertification program.

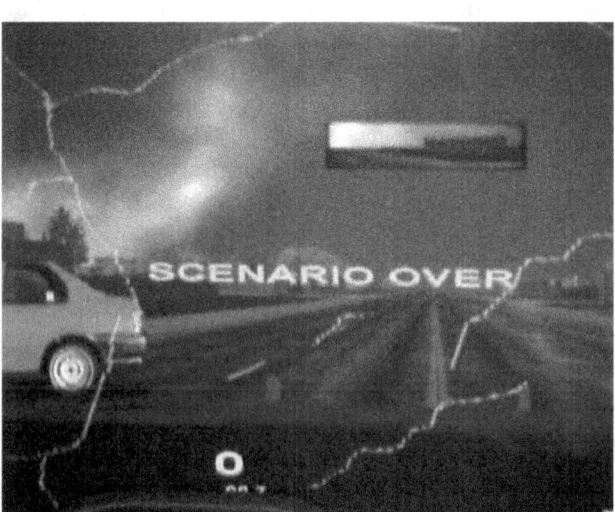

Figure 7-10. Feedback for incorrect decision.

The simulator is part of the driver training offered as a remedial tool through the Sacramento Regional Training Facility. The simulator's software scripting tool allows the instructor to script the scenario of the driver's actual crash. The majority of the time the driver will repeat the same actions he/she took in the actual crash. The instructor then reviews the "drive," discusses the appropriate response(s), and repeats simulations with new scenarios that have the same elements as the original crash.

Effects of the Training
The Sacramento Regional Training Facility was the only agency with data related to training effectiveness. From 1999 to 2003, there was a 77-percent decrease in fire department at-fault accidents. The average net savings to the city and county, after deducting the cost of operating the facility, was $2.5 to $4 per year in litigation alone. This does not include savings related to decreased vehicle maintenance, repair costs, workers' compensation, downtime, etc. In addition to hard data, the instructors receive a great deal of positive feedback from personnel, claiming the training makes them better and safer drivers both on and off duty.

SHARING LESSONS LEARNED
Several organizations involved in this study have developed mechanisms for sharing information on incidents that have occurred. The intent of sharing this information is to prevent future incidents through the recognition of previous bad experiences. The following section highlights some of these information-sharing mechanisms.

Accident Investigation Reporting — The Green Sheet
Any of the following CDFFP incidents triggers an investigation:

- a vehicular accident;
- a severe loss accident; or
- an accident in which there has been a loss of life.

A Multi-Disciplined Accident Investigation Team (MAIT), led by the CHP, conducts the investigation of any significant crashes. Using the MAIT team reduces the potential for conflict of interest. The team consists of five or six members with expertise in the areas related to the incident. For example, the Fleet Regional Manager would be on the investigation team if the incident were equipment-related.

The findings of all investigations are distributed to the entire department through the Intranet using a Green Sheet generated for information purposes. This document gives the incident information, the surrounding circumstances, and the findings of the investigation. The majority of Green Sheets are published within 3 to 4 days. However, if it is a significant and extended investigation, a preliminary sheet may be released, with the final results of the investigation coming in 6 to 8 weeks. Green sheets are used as a training tool and may result in a policy change. An example of a Green Sheet can be found in Appendix H.

National Institute for Occupational Safety and Health (NIOSH) Fatality Investigations
The NIOSH Fire Fighter Fatality Investigation and Prevention Program conducts investigations of both fireground and nonfireground fatal firefighter injuries resulting from a variety of circumstances, including motor vehicle incidents. NIOSH staff also conducts investigations of selected nonfatal injuries. Each investigation results in a report summarizing the incident, and includes recommendations for preventing future similar events. Congress began funding the independent NIOSH investigations in 1998 in response to the need for further efforts to address the continuing national problem of occupational firefighter fatalities. More information and specific investigation results and recommendations are available on the NIOSH Web site at: http://www.cdc.gov/niosh/firehome.html

NIOSH also publishes Hazard ID fact sheets and Alerts related to all aspects of firefighter safety. The NIOSH Hazard ID, *Traffic Hazards to Fire Fighters While Working Along Roadways*, can be found in Appendix H. Other Hazard ID's and Alerts can be accessed online at: http://www.cdc.gov/niosh/othpubs.html

DRIVER REQUIREMENTS

All potential Ventura County Fire Department drivers must complete a task book before being eligible to take the driver's examination. The purpose of this practice is to ensure that potential drivers are familiar with all of the different models of apparatus that they may encounter and the driving course specifications. The tasks that are outlined in the task book are minimum requirements, and are intended to provide a foundation for future learning. This practice encourages motivation and self-direction in members. They must travel to stations that house the apparatus on their own time and make arrangements with the station driver to supervise their task completion. The sections contained in the task book are

- Apparatus Operator;
- Aerial apparatus tasks;
- Type III Wildland Engine tasks;
- Light and air tasks;
- Water Tender tasks;
- Pumping evolution tasks;
- Specialized apparatus tasks;
- Tillered aerial apparatus tasks; and
- Aircraft Crash Vehicle tasks.

The completed book is submitted to the Training Section, where it is reviewed and recorded in the training database. The original book is placed in the individual's training file. Examples of driving course specifications from the task book can be found in Appendix I.

EMERGENCY VEHICLE MAINTENANCE

Each year a percentage of firefighter/EMS injuries and deaths are the result of mechanical problems and apparatus failure. In these cases, the public and legal systems are increasingly inclined to look to the organization's adherence to maintenance standards. In an ongoing attempt to minimize risks and fatalities, several groups have been working together to establish and implement standards, training and education, and certification programs supporting the safety of emergency vehicle equipment.

In August of 2000, NFPA 1071, *Standard for Emergency Vehicle Technician Professional Qualifications*, was issued. This standard establishes a set of professional qualifications that can be used to develop educational requirements and corresponding certifications for emergency vehicle technicians and mechanics. In addition, NFPA 1915, *Standard for Fire Apparatus Preventative Maintenance Program*, provides guidance for creating and maintaining a comprehensive maintenance program. Together, these standards can be used to ensure that a department's staff has skills adequate to service and maintain the full spectrum of emergency vehicles. Although NFPA standards are not legally binding unless formally adopted by the authority having jurisdiction (AHJ), many departments, companies servicing emergency equipment, and original equipment manufacturers have adopted NFPA 1071 and NFPA 1915 as part of their internal policies and operating procedures.

In an ongoing effort to ensure vehicle safety, the Emergency Vehicle Technician (EVT) Certification Commission was established to write and administer tests that would demonstrate proficiency in established standards. The tests resemble those used by the Automotive Service Excellence organization (ASE), applying their high "blue seal of excellence" standards to fire equipment. Technicians who receive all of the EVT and ASE certifications are recognized as a master certified EVT. The EVT certification program presently has two certification tracks, one for technicians who service and maintain fire apparatus and another for technicians who service and maintain ambulances. The levels of fire apparatus certification are shown in Table 7-2. The levels of ambulance certification are shown in Table 7-3.

TABLE 7-2.
EMERGENCY VEHICLE TECHNICIAN FIRE APPARATUS CERTIFICATIONS

Fire Apparatus Technician Level Requirements – Level I
ASE Exams:	EVT Exam:
• T-7 or A-7, Heating and Air Conditioning • T2 Truck, Diesel Engines • T4 Truck, Brakes • T5 Truck, Suspension and Steering	• F2 Design & Performance Standards and Preventive Maintenance of Fire Apparatus

Fire Apparatus Technician Level Requirements – Level II
ASE Exams:	EVT Exams:
• T3 Truck, Drive Train • T6 Truck, Electrical Systems	• F3 Fire Pumps & Accessories • F4 Fire Apparatus Electrical Systems

Fire Apparatus Technician Level Requirements – Master Level III
ASE Exams:	EVT Exams:
• T1 Truck, Gasoline Engines	• F5 Aerial Fire Apparatus • F6 Allison Automatic Transmissions

---

TABLE 7-3.
EMERGENCY VEHICLE TECHNICIAN AMBULANCE CERTIFICATIONS

Ambulance Technician Level Requirements – Level I
ASE Exams:	EVT Exams:
• A-4 Automobile, Suspension and Steering • A-5 Automobile, Brakes • A-6 Automobile, Electrical Systems • A-8 Automobile, Engine Performance	• E-1 Design & Performance and Preventive Maintenance of Ambulances

Ambulance Technician Level Requirements – Level II
ASE Exams:	EVT Exams:
• A-1 Automobile, Engine Repair • A-3 Automobile, Manual Drive Train & Axle • A-7 Automobile, Heating and Air-Conditioning • T-2 Truck, Diesel Engines	• E-2 Ambulance Electrical Systems • E-3 Ambulance Heating, Air-Conditioning, & Ventilation

Ambulance Technician Level Requirements – Master Level III
ASE Exams:	EVT Exams:
• A-2 Automobile, Automatic Transmission and Transaxle • T-4 Truck, Brakes • T-5 Truck, Suspension and Steering	• E-4 Ambulance Cab, Chassis, and Body

Today, there are approximately 60,000 emergency vehicle technicians and mechanics in the United States. It is estimated that fewer than 25 percent have received certification meeting the requirements outlined in NFPA 1071. It is a common belief that increasing the number of certified emergency vehicle technicians will assist in reducing the number of emergency worker injuries and fatalities related to equipment failure. Ensuring that quality educational opportunities are readily available is central to increasing the number of certified EVT's.

More information about the EVT Certification Commission is available online at:

http://www.evtcc.org

Additional information about the ASE is available online at http://www.ase.com

RECOMMENDATIONS FOR TRAINING

- Require driver training instructors to possess appropriate instructor credentials and firsthand experience in emergency vehicle operations.

- Include an appropriate blend of theory, simulation, and hands-on practice in all driver training programs.

- Limit the use of a driving simulator in training programs to improving drivers' decisionmaking processes.

- Require all emergency vehicle drivers to participate in refresher training on an annual basis and recertify according to department requirements no less than every 3 years.

- Require members to be familiar with all of the different models of fire apparatus that they may be expected to operate before taking the driver's examination.

- Share the findings of all investigations with the entire department in a timely manner.

---

[1] Zagaroli, L. & Taylor, A. (2003, January 1). Ambulance driver fatigue a danger. *Detroit News*.

SUMMARY OF RECOMMENDATIONS

INTRODUCTION

Since 1984, motor vehicle crashes have accounted for 20 to 25 percent of firefighter fatalities annually. Between 1990 and 2000, 18 percent of the fatalities occurred responding to an alarm, and 4.1 percent occurred returning from an alarm. Of the firefighters who died in motor vehicle crashes, 25 percent were killed in privatelyowned vehicles (POV's). Following POV's, the apparatus most often involved in fatal collisions were tankers, engines, and airplanes. Tankers claimed more fatalities than engines and aerial apparatus combined. Approximately 27 percent of fatalities were ejected at the time of the collisions. Reportedly, only 21 percent were wearing restraints prior to the crash. EMS personnel in the United States have an estimated fatality rate of 12.7 per 100,000 workers. As traffic volume increases and the highway and interstate system becomes more complex, emergency responders face a growing risk to personal safety while managing and working at highway incidents.

This report identified various progressive safety practices and programs related to emergency vehicle operations. Although all of the information in this report is pertinent, some factors stand out as being especially relevant. The purpose of this report is to identify practices with potential for improving member safety when responding to and returning from a response and during highway operations. This chapter provides a summary of recommendations based on suggestions of forum participants and the most important practices discussed throughout the report.

Develop a comprehensive database that tracks accidents involving emergency vehicles and any resulting injuries/deaths to both firefighters and civilians.
No solid data are available to support any of the progressive programs aimed at improving member safety related to vehicle and highway operations. There should be a national repository that collects data from all organizations. Because data are so fragmented and minimal, there is only anecdotal information to support the effectiveness of these safety practices or programs.

Mark apparatus with conspicuous, contrasting colors.
Striping with Diamond Grade fluorescent material in a herringbone pattern, such as that used by the Plano, Texas, Fire Department improves driver recognition while the apparatus is on the roadways.

Consider visibility and conspicuity when designing color and placement of additional warning lights on vehicles.
These factors are not highly dependent on driver behavior. For example, the Plano Fire Department uses additional lighting of a contrasting color to the vehicles.

Install contrasting colored restraints and a rearview mirror above the officer's seat.
NFPA 1901 states, "If available from the chassis manufacturer, the seatbelt webbing shall be bright red in color." The committee suggested that "International Orange" is not the best color because it shows dirt. A rearview mirror above the officer's seat allows him/her to assure that all personnel are wearing restraints without the distraction of having to turn around. These practices make it easier to assure that all personnel are restrained before the apparatus moves.

Use spotters when backing the apparatus.
Even though cameras and other devices that assist with backing the apparatus do provide some measure of safety, there is no substitute for having at least one, preferably two, spotters to guide the driver while the apparatus is being operated in reverse. NFPA 1500 requires spotters for backing, regardless of whether the apparatus is equipped with cameras or other backing safety equipment. One spotter should be equipped with a portable radio in the event that he/she needs to contact the driver during the backing operation.

Consider optical preemption to improve emergency vehicle movement through controlled intersections.
With activation at 1/2 mile, optical preemption usually clears the intersection before the unit arrives, allowing continued travel without the need to stop. However, drivers must not assume they will have a clear path when they arrive at the intersection and must verify visually that all traffic has stopped.

Always bring units to a complete stop at red lights, stop signs, and activated or unguarded rail crossings before proceeding.
Even though optical preemption devices can provide a clear path for emergency units through traffic signals, drivers must not anticipate a change so far ahead that they cannot stop the apparatus safely or avoid striking another vehicle.

Pursue a working relationship with the State Department of Transportation and local law enforcement to identify criteria for responses that would incorporate DOT resources to aid in traffic control and improve safety for responders involved in highway operations.
An intelligent transportation system, traffic operations center, and DOT freeway motorist assist patrollers, if available, can provide considerable assistance to emergency response departments for traffic control at a highway operations scene.

Position the engine at a 45-degree angle to the lanes with the pump panel toward the incident and the front wheels rotated away from the incident when conducting highway operations.
In the event that a motorist strikes the engine, the engine will act as a barrier and in the unlikely event the engine is moved upon impact, it will travel away from the work zone. The pump panel should face the incident to provide protection for the operator while monitoring apparatus functions.

Extinguish forward-facing emergency vehicle lighting, especially on divided roadways.
This will help reduce distractions and glare to oncoming drivers. The headlights on the apparatus can temporarily blind approaching drivers, resulting in the problem of glare recovery. It takes at least 6 seconds, going from light to dark, and 3 seconds from dark to light for vision to recover.

Create a safe work environment.
Reduce the use of lighting as much as possible at the scene. Establish good traffic control, including placement of advance warning signs and traffic control devices to divert or detour traffic. This will allow responders to reduce emergency vehicle lighting. This is especially true for major traffic incidents that might involve a number of emergency vehicles.

Require members to wear highly reflective material when conducting highway operations.
Personnel visibility is critical during highway operations. In addition to PPE, high visibility vests should meet the ANSI III Standard. ANSI III provides the highest level of both retroreflective and florescent properties. It makes workers conspicuous through a full range of body movements at 1,280 feet.

Remain vigilant during all phases of highway operations.
Even with all safety precautions in place, personnel are at risk from drivers who may violate safety zones, blocking apparatus, cones, curbs, etc. Use law enforcement personnel to help secure and protect the scene. If necessary, appoint a Safety Officer to monitor traffic.

Work with neighboring districts to develop similar highway operations policies.
Mutual-aid departments using the same apparatus positioning and marking policies will improve the overall safety of all members at the incident, through understanding and familiarity with operating procedures.

Allow all members to submit suggested policy changes.
Those members who work at highway operations on a daily basis can provide a critical review of existing policy and valuable information to improve safety based on experience.

Assess and code responses using the classic risk management matrix. Identify the low-frequency, low-severity incidents, use preset dispatch questions, and respond in a nonemergency mode.
Multiple EMS studies have shown, with the exception of a few conditions, that there is no statistically significant difference in patient outcomes with emergency driving on ambulances versus nonemergency driving. It is logical to assume that in many situations (manpower for EMS, water leak, etc.) the results related to incident outcome would be duplicated with fire apparatus. St. Louis Fire Department found responding without lights and sirens reduced its crash rate. There are commercial dispatch programs available for priority dispatching; or agencies can develop their own specific questions, similar to those developed by the Salt Lake Valley, Utah, departments.

Require driver training instructors to possess appropriate instructor credentials and firsthand experience in emergency vehicle operations.
Organizations that deliver driver training should ensure that their instructors meet local and State certification requirements for fire instructors. These instructors should also have specialized training and experience in safe and proper operation of emergency vehicles.

Include an appropriate blend of theory, simulation, and hands-on practice in all driver training programs.
Students should be provided with the theory and concepts associated with the safe operation of emergency vehicles and highway emergency scene safety. If driving simulators are available, they should be used to familiarize driver candidates with principles of vehicle operation. Tabletop scenarios also can be used for simulations. Practical driving exercises should be conducted on a controlled driving course and over public thoroughfares.

Limit the use of a driving simulator in a training program to improving drivers' decisionmaking processes.
A driving simulator is an effective tool to familiarize driver candidates with vehicle operations. It can also be used to place experienced drivers in special situations that cannot be reproduced practically or safely using actual vehicles. It is most appropriate for high-risk situations, such as intersections and conflict management. It also allows remedial training to be tailored to a specific problem. However, a simulator cannot replace practical driver training exercises using actual emergency vehicles. All driver training programs must include the operation of actual vehicles in order to ensure candidates are capable of operating the real thing.

Require all emergency vehicle drivers to participate in refresher training on an annual basis, and recertify according to department requirements no less than every 3 years. The organization's in-service training program must stress continuous maintenance and improvement of driver skills. In order to ensure that emergency vehicle drivers maintain their skills at an acceptable level, they should be required to prove their proficiency by recertifying every 3 years.

Require members to be familiar with all of the different models of fire apparatus that they may be expected to operate before taking the driver's examination.
A member who is going to drive and operate fire department apparatus must be familiar with all the different types of fire apparatus operated by the department. One method of developing this familiarity is to require the member to complete a task book before taking the driver's examination. This practice assures the member is familiar with all of the different models of apparatus that he/she may encounter and any driving course specifications. The tasks should be minimum requirements and provide a foundation for future learning.

Share the findings of all investigations with the entire department in a timely manner. Sharing incident information, the surrounding circumstances, and the findings of the investigation in a timely manner can be a valuable training tool. If the investigation is prolonged, release a preliminary findings report.

REFERENCES

Addario, M., Brown, L., Hogue, T., Hunt R. C., & Whitney, C. L. (2000). Do warning lights and sirens reduce ambulance response times? *Prehospital Emergency Care,* *4*(1), 70-74.

Allen, M., Strickland, J., & Adams, A. (1967). As cited in DeLorenzo, R. & Eilers, M. (1991). Lights and siren: A review of emergency vehicle warning systems. *Annals of Emergency Medicine,* online.

Davis, C. C., (1982). *Accidents involving stopped vehicles on freeway shoulders.* Automobile Club of Southern California.

Davis, R. (2001, March 21). Speeding to the rescue can have deadly results. *USA Today.*

DeLorenzo, R. & Eilers, M. (1991). Lights and siren: A review of emergency vehicle warning systems. *Annals of Emergency Medicine,* online.

Federal Emergency Management Agency, U.S. Fire Administration. (1994-2001). *Fire Fighter Fatality in the United States.*

Harvey, N. (2003, May 25). Service honors fallen EMS providers. *Roanoke Times.*

Hills, B.L. (1980). As cited in DeLorenzo, R. & Eilers, M. (1991). Lights and siren: A review of emergency vehicle warning systems. *Annals of Emergency Medicine,* online.

Kahn, C.A., Pirrallo, R.G., & Kuhn, E.M. (2001). Characteristics of fatal ambulance crashes in the United States: an 11-year retrospective analysis. *Prehospital Emergency Care, 5,* 261-269.

Kupas, D., Dula, D., & Pino, B. (1994). Patient outcome using medical protocol to limit "lights and siren" transport. *Prehospital and Disaster Medicine. 9*(4).

National Bureau of Standards. (1978). *Emergency vehicle warning lights: state of the art,* Pub. No.480-16.

National Fire Protection Association. (2003). NFPA 1901: *Standard for automotive fire apparatus.* Quincy, MA: NFPA.

- (2002). NFPA 1041: *Standard for fire service instructor professional qualifications.* Quincy, MA: NFPA.

- (2002b). NFPA 1500: *Standard on fire department occupational safety and health program.* Quincy, MA: NFPA.

- (2000). NFPA 1071: *Standard for emergency vehicle technician professional qualifications.* Quincy, MA: NFPA.

- (2000b). NFPA 1915: *Standard for fire apparatus preventative maintenance program.* Quincy, MA: NFPA.

Scarano, S. (1981). As cited in DeLorenzo, R. & Eilers, M. (1991). Lights and siren: A review of emergency vehicle warning systems. *Annals of Emergency Medicine,* online.

Solomon, S. (2002). The case for amber emergency warning lights. *Firehouse, 27*(2), 100-102.

U.S. Department of Health and Human Services, Centers for Disease Control and Prevention. (2003, February 28). Ambulance crash-related injuries among emergency medical services workers — United States, 1991-2002. *Morbidity Mortality Weekly Report, 52* (08), 154-156.

U.S. Department of Health and Human Services, National Institute for Occupational Safety and Health. (2001). 26-year-old emergency medical technician dies in multiple fatality ambulance crash—Kentucky. Pub. No. FACE-2001-11.

Zagaroli, L. & Taylor, A. (2003, January 1). Ambulance driver fatigue a danger. *Detroit News.*

PUBLIC COMMENTS ON NFPA 1901

Standard on Automative Fire Apparatus
SUBMITTED BY THE EVSI
FORUM PARTICIPANTS

Comment #1
Recommended wording from EVSI:

Add a new paragraph as 12.1.4 to read as follows:

A high visibility plate stating the overall vehicle height, weight (loaded or GVWR) and length shall be permanently attached in close proximity to the transmission shift lever (or pad).

Wording that went into the requirements of NFPA 1901:

12.1.4 The fire apparatus manufacturer shall permanently affix a high visibility plate in a location visible to the driver while seated.

12.1.4.1* The plate shall show the height of the completed fire apparatus in feet and inches or meters, the length of the completed fire apparatus in feet and inches or meters, and the gross vehicle weight rating (GVWR) in pounds or kilograms.

12.1.4.2 Wording on the plate shall indicate that the information shown was current when the apparatus was manufactured and that, if the overall height changes while the vehicle is in service, the fire department must revise that dimension on the plate.

Wording that went into the annex of NFPA 1901:

A.12.1.4.1 It is important for fire apparatus drivers to understand the height, length and weight of the vehicle compared to their personally owned vehicles. It is also important that this information be accurate. Because the height of the apparatus could change after delivery, depending on what equipment might be added, the fire department must note such changes on the plate. Suggested wording for the plate is shown in Figure A.12.1.4.1

Figure A.12.1.4.1 Suggested Plate Showing Dimensions of Fire Apparatus.

```
When manufactured, this vehicle was:
          XX ft YY in. High
          XX ft YY in. Long
          ZZZZ lbs GVWR
   Changes in height since the apparatus was
              manufactured
shall be noted on this plate by the fire department.
```

Committee substantiation for its changes:

The committee agrees with the intent of the submitter. However, it feels that the location should be left to the manufacturer as long as it is visible to the driver when he/she is seated. The committee has rewritten the submitters requirement to be more specific and added an annex item to assist users with understanding the requirement. The committee also notes that it is important the fire departments note any changes on the plate that change the height of the vehicle after it is delivered.

Comment #2
Recommended wording from EVSI:
Revise 14.1.3 of the draft to read as follows:

Each crew riding position shall be provided with a seat and an approved seat belt designed to accommodate a person with and without heavy clothing. Seat belt webbing and the "buckles" shall be high visibility "International Orange" in color with the releasing "button" being a contrasting color. The buckle portion of the seat belt shall be mounted on a rigid or semi-rigid stalk such that the buckle remains positioned in a convenient location as would allow the fastening of the belt to be an operation capable of being completed using only one hand.

Wording that went into the requirements of NFPA 1901:
14.1.3.1 If available from the chassis manufacturer, the seat belt webbing shall be bright red in color and the buckle portion of the seat belt shall be mounted on a rigid or semi-rigid stalk such that the buckle remains positioned in an accessible location.

Committee substantiation for its changes:
The committee feels "International Orange" is not the best color because it shows dirt and recommends "red" for seat belts. It is adding wording concerning the provision of a rigid or semi-rigid stock for mounting the buckle.

Comment #3
Recommended wording from EVSI:
Add a new paragraph as 14.1.6 to read as follows:

All crew and driving compartment doors shall have at least 96 square inches of reflective material or a reflective "Stop Sign" affixed to the inside of each of the doors.

Wording that went into the requirements of NFPA 1901:
14.1.6 All driving and crew compartment doors shall have at least 96 in^2 (62,000 mm^2) of reflective material affixed to the inside of each of the doors.

Committee substantiation for its changes:
The Committee agrees with the comment except for the "Stop Sign" option. The "Stop Sign" option could cause confusion to motorists. The committee also feels that often it is neither necessary nor desired for the motorist to stop.

Comment #4
Recommended wording from EVSI:
Revise 14.3.3 in the draft to read as follows:

"The passenger side mirror shall be remote controlled from the drivers position and be so mounted that the driver has a clear view of the mirror..."
Wording that went into the requirements of NFPA 1901:
None

Wording that went into the annex of NFPA 1901:
A.14.3.3 (par. 2) When specifying new apparatus, the purchaser should consider remotely controlled mirrors, especially on the passenger side. The location and mounting of the mirrors should not be placed where door pillars or other obstructions block their view. The location and mounting should be placed so warning lights do not reflect in the mirror to blind the driver's view. The location and mounting should not be placed so that the driver must look through the windshield area that is not wiped by the windshield wiper when viewing the passenger side mirror. Convex and other secondary mirrors should be considered to eliminate blind spots not covered by primary mirrors. Where necessary, heated mirrors should also be considered.

NFPA Staff note: This annex material is the result of action on 2 public comments combining new wording with existing annex wording.

Committee substantiation for its changes:
The committee thinks the proposed requirement could cause reliability problems and should not be a requirement. However, the committee thinks it is appropriate to place the intent of the submitter's recommendation in the annex for fire departments to consider.

Comment #5
Recommended wording from EVSI:
Add a new paragraph as 15.9.4 to read as follows:

Where body compartment doors are of "roll-up" construction, a strip of red and white reflective material (like what is currently placed on all highway trailers) meeting the current D.O.T. Standard, shall be affixed in the rub rail area below the door. If compartment doors are of the "swing-out" type, a 4 in. (minimum) reflective stripes or "Chevron type" reflective stripes shall be placed on the inside of the doors.

Wording that went into the requirements of NFPA 1901:
None

Wording that went into the annex of NFPA 1901:
A.15.9.3.2 If fire departments specify rollup doors, they should consider affixing a strip of reflective material to the rail area below the door. If fire departments specify vertically hinged compartment doors, they should consider affixing 4 in. (100 mm) minimum width reflective stripes or chevron-type reflective stripes on the inside of the doors.

Committee substantiation for its changes:
The committee thinks the intent of this comment is appropriate for the annex rather than the main body of the standard. Also, this comment does not take into consideration how the minimum reflective material will be placed on a horizontally hinged door.

UDOT LETTER OF AGREEMENT FOR OPTICAL PREEMPTION

Date:_____

Chief _____

_____ Fire Department

_____

_____

Re: Traffic Signal Pre-emption

Dear Chief _____,

The Utah Department of Transportation approves your request to install traffic signal preemption on UDOT-owned traffic signals at the following locations in _____

_____

_____

The following conditions apply:

1. UDOT approves the use of traffic signal pre-emption as a public safety measure at locations where local Fire Departments find it necessary to safely move emergency vehicles through high traffic volume intersections under conditions of emergency response. Accordingly, the _____ Fire Department agrees to install vehicle emitters only on Fire Department vehicles and to use traffic signal pre-emption only under emergency conditions. _____ Fire Department policies shall be modified or updated if necessary to limit the use of traffic signal pre-emption to appropriate emergency conditions.

2. UDOT will provide technical advice in design, installation, and maintenance at no expense to the _____ Fire Department.

3. The _____ Fire Department shall bear all costs of the equipment and the labor for a complete installation. Such costs will include the replacement or upgrading of any existing conduit or cabinets which are inadequate to accommodate the new pre-emption equipment.

4. The _____ Fire Department may select a contractor of its choice, provided that the contractor is pre-qualified by UDOT for traffic signal work.

5. After installation, UDOT will provide routine maintenance such as cleaning and aiming of receivers, and maintenance of the proper programming of the equipment. _____ shall report any malfunctions requiring UDOT assistance to the UDOT Traffic Operations Center at 887-3700. The Fire Department will be responsible for all maintenance of the emitters.

6. The _____ Fire Department will be responsible for replacing any malfunctioning equipment, if necessary, in the future if it cannot be routinely repaired by UDOT.

7. UDOT's policy is to use traffic signal equipment which provides standardized interfaces and open communication protocols, thus permitting compatibility and interchangeability between equipment provided by different vendors. At this time, the pre-emption devices commonly available on the market do not fully meet this objective. Therefore, UDOT does not have an approved equipment

list, nor does it endorse or specify any particular brand or vendor. UDOT will approve equipment proposed for use by local governments Departments on a case-by-case basis. At this time, UDOT will approve equipment that has:

- Proven reliability
- A means of uniquely identifying and logging pre-emption calls by individual vehicles.
- Sufficient precision and accuracy to minimize false pre-emption calls which are disruptive to normal operation of the traffic signal.

8. A proposed design (sketch plan), equipment list, and proposed contractor must be submitted for approval by UDOT prior to start of installation work.

9. Vehicle emitter shall be wired so that they are enabled only when emergency light bars on the vehicles are activated. A means shall be provided to automatically disable the emitters when the vehicle is parked.

10. Upon completion of the work, the _____ Fire Department must arrange for the manufacturer's representative to be on-site at no cost to UDOT to inspect the final installation, and assist in testing, fine-tuning and programming of the pre-emption unit. The manufacturer must provide a letter to UDOT certifying that the pre-emption unit and the vehicle emitters have been installed in accordance with the manufacturer's recommendations and with UDOT requirements.

APPROVED:

_____   _____
David A. Kinnecom, P.E.   Date
Traffic Operations Engineer
Utah Department of Transportation

_____   _____
Chief   Date
_____ Fire Department

CC:

Part 1: Fairfax County Fire and Rescue
Operating Procedures for Highway Incidents

Part 2: Phoenix Fire Department
Safe Parking While Operating in or Near Vehicle Traffic

FAIRFAX COUNTY FIRE AND RESCUE
OPERATING PROCEDURES FOR HIGHWAY INCIDENTS
PREFACE

The primary objectives for any operation at the scene of a highway incident are preserving life, preventing injury to emergency workers, protecting property and restoration of traffic flow.

Managing a highway incident and other related problems is a team effort. Each responding agency has a role to play in an effective incident operation. The Police, Virginia Department of Transportation and the Fire and Rescue Department all play important roles in the management of highway incidents. It is not a question of "Who is in charge?" but "Who is in charge of <u>what</u>?"

Care of the injured, protection of the public, safety of the emergency responders and clearance of the traffic lanes should all be priority concerns of the incident manager operating at the scene of a highway accident. It is extremely important that all activities that block traffic lanes be concluded as quickly as possible and the flow of traffic be allowed to resume promptly. When traffic flow is heavy, a small savings in accident scene clearance time can greatly reduce traffic backups and reduce the probability of a secondary incident. Restoring the roadway to normal or to as near normal as soon as possible creates a safer environment for the motorist and emergency responders. Additionally, it improves the public's perception of the agencies involved and reduces the time and dollar loss resulting from the incident.

The primary objectives for any operation at the scene of a highway incident are preserving life, preventing injury to emergency workers, protecting property and restoration of traffic flow.

PURPOSE

The purpose of this manual is to provide the incident officers and members of the Fire and Rescue Department with a uniform guide for safe operations at incidents occurring on the highway system. It is intended to serve as a guideline for decision-making and can be modified by the incident officers as necessary to address existing incident conditions.

The most common occurrence and the one that possibly has the greatest potential for an unfavorable outcome to department personnel are emergency operations at the scene of a vehicle accident. Each year many significant incidents occur on roadways within Fairfax County. Whether it is on the interstate highway or on a secondary road, the potential for injury or death to a member of the department is overwhelming.

RESPONSE

Fire and Rescue department personnel shall operate safely and make every effort to minimize the risk of injury to themselves and those who use the highway system. Personnel shall wear appropriate gear and be seated with restraint strap on prior to their vehicle responding to all incidents.

Apparatus operating in the emergency mode shall operate warning devices and follow the guidelines outlined in Department S.O.P. 1.8.03.

WARNING LIGHTS:	Emergency warning lights shall remain operational while responding and, when necessary while working at incidents.
HEADLIGHTS:	Apparatus headlights shall be operational during all responses and incidents regardless of the time of the day. Caution should be used to avoid blinding oncoming traffic while on the scene.
SIREN AND AIR HORN:	When operating as an emergency vehicle, siren and air horn will be utilized.

Emergency response to incidents on limited access highways should include at least one unit traveling each direction on the highway. When units respond together in the same direction, they should remain in single file in relatively close proximity to one another. This reduces confusion to the motorists on the highway as to how to appropriately yield the right of way to emergency apparatus.

The preferable lane of response should be the left travel lane. When the shoulder must be utilized, apparatus operators must use extreme caution. Be aware of road signs, debris, guard rails, oversized vehicles and stopped vehicles. Fire and Rescue Department vehicle operators must reduce speed of their vehicle when using the shoulder of the road to access the incident.

Response on access ramps shall be in the normal direction of travel, unless an officer on the scene can confirm that oncoming traffic has been stopped and no civilian vehicles will be encountered on the ramp.

Median strip crossovers marked "Authorized Vehicles Only" shall only be used for turning around and crossing to the other travel lanes, when apparatus can complete the turn without obstructing the flow of traffic in either travel direction or all traffic movement has stopped. Under no circumstances shall crossovers be utilized for routine changes in travel direction.

Utilization of U-turn access points in "Jersey" barriers on limited access highways is extremely hazardous and shall be utilized only when the situation is necessary for immediate lifesaving measures.

ON SCENE ACTIONS

The proper spotting and placement of apparatus is the joint responsibility of the driver and officer. The proper positioning of apparatus at the scene of an incident assures other responding resources of easy access, a safe working area and helps to contribute to an effective overall operation.

The unit officer is responsible for the safety of the unit and his/her crew from the time the apparatus leaves quarters until its return. Safety of the crew is foremost while they are operating, both in emergency and non-emergency situations.

ARRIVAL

Standard practice shall be to position apparatus in such a manner as to ensure a safe work area at least one lane wider than the width of the incident. This may be difficult to accomplish at incidents on secondary and one-lane roads. Position the apparatus in such a manner as to provide the safest work area possible.

A work zone shall be established allowing EMS units and the Rescue Squad to position in close proximity of the incident (Figure 1).

The Engine placement should be back some distance from the incident, utilizing it as a safety shield blocking only those travel lanes necessary. The Engine shall be placed at an angle to the lanes, with the pump panel toward the incident and the front wheels rotated away from the incident.

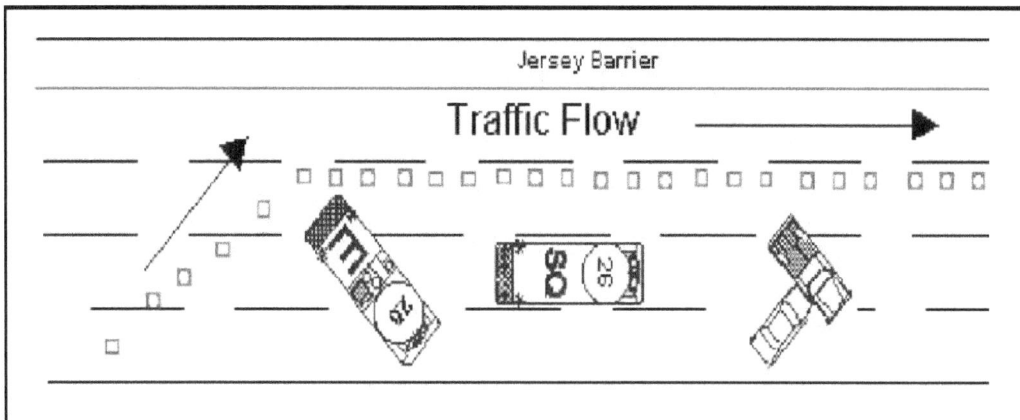

Figure 1.

In the event that a motorist strikes the engine, the engine will act as a barrier and in the unlikely event the engine is moved upon impact, it will travel away from the work zone. The pump panel should face the incident to provide protection for the operator while monitoring apparatus functions.

Before exiting apparatus at an incident, personnel shall check to ensure that traffic has stopped to avoid the possibility of being struck by a passing vehicle. Personnel should remember, to look down to ensure that debris on the roadway will not become an obstacle, resulting in a personal injury. All crewmembers shall be in full protective clothing or traffic vests as the situation indicates.

As soon a possible, the engine operator should place out flares and traffic cones. Traffic cones assist in channeling traffic away from the incident. Cones shall be used whenever department vehicles are parked on or about any road surface.

Placement of cones and/or flares shall begin closest to the incident, working towards on-coming traffic. Cones and/or flares shall be placed diagonally across the roadway and around the incident. This assists in establishing a safe work zone. When placing cones or flares, care should be exercised to avoid being struck by oncoming traffic (Figure 2).

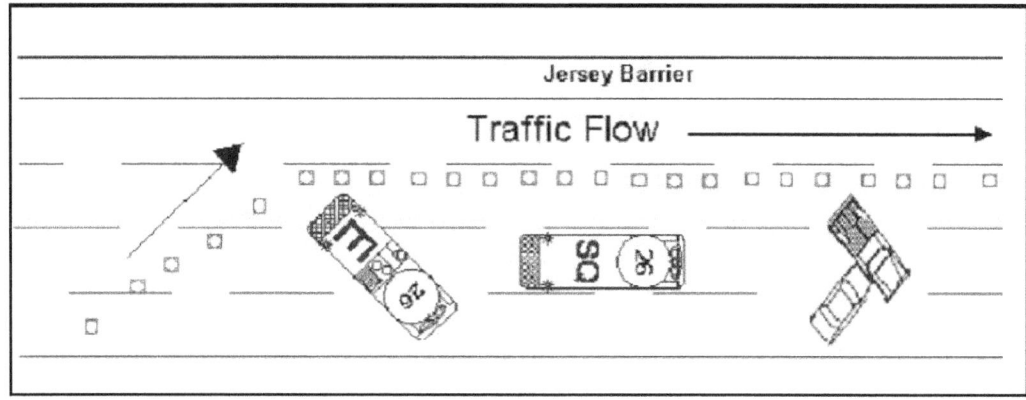

Figure 2.

The speed of traffic must be considered when establishing a safe work area.

Utilize the following chart to determine how far to place the first cone or flare away from the incident scene.

POSTED SPEED LIMIT	DISTANCE
35 MPH	100 Ft.
45 MPH	150 Ft.
55 MPH	200 Ft.
>55 MPH	250 Ft. plus...

When channeling traffic around the incident, cones shall also be used in front of the incident with the same diagonal placement to direct traffic safely around the work zone (Figure 3).

It is possible to channel traffic around a curve, hill or ramp provided the first cone is placed such that the oncoming driver is made aware of imminent danger. The first cone should be placed well before the curve, hill or ramp. The rest of the cones shall be placed diagonally across the lanes around the work zone.

A Four Point System will be used whenever vehicles are parked in an area which does not require the channeling of traffic. One cone will be placed at each corner of the vehicle approximately four feet from each corner. This will assist the motorist and incoming units to identify the established work zone. When utilizing this system around aerial apparatus and rescue squads, additional cones should be placed to identify extended outriggers, booms and heavy equipment (Figure 4).

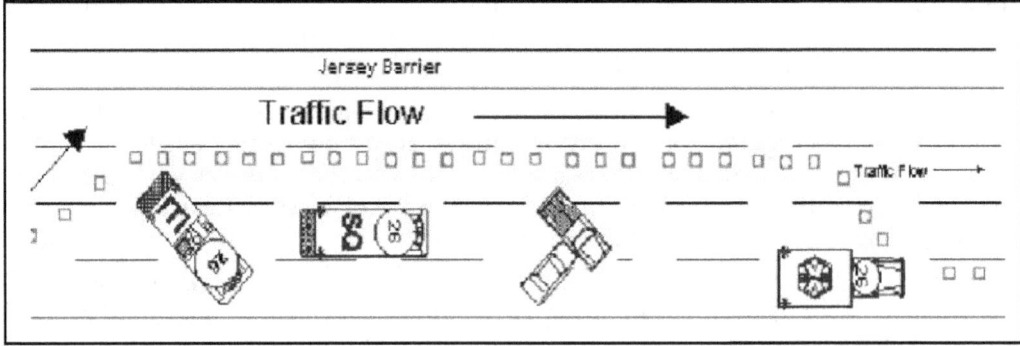

Figure 3.

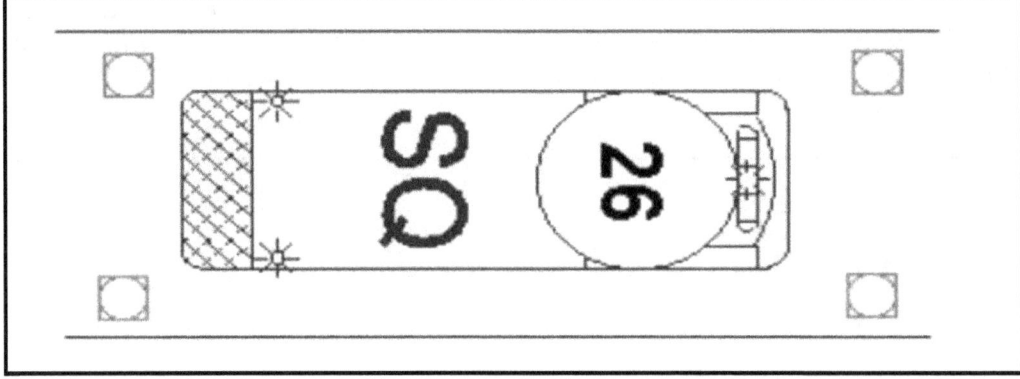

Figure 4.

PARKING OF RESPONSE VEHICLES

Providing a safe work area for emergency responders is a priority at every emergency incident. However, consideration must be given to keeping as many traffic lanes as possible open. Except for those vehicles needed in the operation and those used, as a shield for the work area, other response vehicles should be parked together in a designated area. Parking should be on the shoulder or median area, if one exists. Parking response vehicles completely out of available travel lanes greatly assists in the movement of traffic. If not needed to illuminate the scene, drivers should remember to turn vehicle headlights off when parked at incidents.

APPARATUS VISIBILITY AT NIGHT

As the human eye becomes adapted to the dark, the first color to leave the spectrum is red. This is important due to the fact that our warning lights are red. The color red tends to blend in to the nighttime surroundings.

Glare vision and recovery is the amount of time required to recover from the effects of glare once a light source is passed through the eye. This takes at least six seconds, going from light to dark and three seconds from dark to light for vision to recover.

At 50 miles per hour, the distance traveled during a second is approximately 75 feet. Thus, in six seconds, the vehicle has traveled 450 feet before the driver has fully regained night vision. This is extremely important when operating on roadways at night.

The headlights on the apparatus can temporarily blind drivers that are approaching fire and rescue apparatus. Drivers of oncoming vehicles will experience the problem of glare recovery. This essentially means individuals are driving by the emergency scene blind. The wearing of protective clothing and or traffic vests will not help the blinded driver see department members standing in the roadway.

Studies conducted show that at two and a half car lengths away from a vehicle with its headlights on, the opposing driver is completely blinded. The best combination of lights to provide maximum visibility is as follows:
Red warning lights on
Headlights off
Fog lights off
Pump panel lights on
Spot lights on rear (and front if equipped) on and directed on to a traffic cone.
Traffic directional boards operating

Low beam headlights can be utilized to light the emergency scene using care as to light only the immediate scene.

CLEARING TRAFFIC LANES

When outside of a vehicle on a major roadway, both civilian and emergency responders are in a very unsafe environment. Therefore, it is imperative to take every precaution to protect ourselves as well as civilians at incident scenes. Positioning apparatus to serve as a shield for work areas is a prudent practice on any incident on a major roadway. But, we must remember that reducing and/or shutting down traffic lanes creates other problems and safety concerns. Therefore, it is critical when operational phases (extrication's, medical care and suppression) are completed, that apparatus be repositioned to allow traffic to flow on as many lanes as possible.

Remember that unnecessarily closing or keeping traffic lanes closed greatly increases the risk of a secondary incident occurring in the resulting traffic backup. One minute of stopped traffic causes an additional four-minute delay in traffic.

CLEARING THE SCENE

Management of incidents on the interstate system and local roadways requires the expertise and resources of the Fire and Rescue Department, Police Department, Virginia State Police, local Police Departments, and the Virginia Department of Transportation working in concert. While the safety of emergency services personnel is the paramount concern for the officer in charge, the flow of traffic must be kept in consideration at all times. The closing of roadways disrupts traffic throughout the area as well as having a significant impact on businesses throughout the region.

Keeping the safety of all personnel in mind and coordinating the needs with the other emergency services, the officer in charge should begin to open closed lanes utilized for extrication and place units in service as soon as practical.

PHOENIX FIRE DEPARTMENT
SAFE PARKING WHILE OPERATING IN OR NEAR VEHICLE TRAFFIC
M.P. 205.07A04/95-R

RETURN TO VOLUME 2 INDEX
- OVERVIEW
- SAFETY BENCHMARKS
- FREEWAY OPERATIONS

OVERVIEW

This procedure identifies parking practices for fire department apparatus that will provide maximum protection and safety for personnel operating in or near moving vehicle traffic. It also identifies several approaches for individual practices to keep firefighters safe while exposed to vehicle traffic.

IT SHALL BE THE POLICY OF THE PHOENIX FIRE DEPARTMENT TO POSITION APPARATUS AT THE SCENE OF EMERGENCIES IN A MANNER THAT BEST PROTECTS THE WORK AREA AND PERSONNEL FROM VEHICLE TRAFFIC AND OTHER HAZARDS.

All personnel should understand and appreciate the high risk that firefighters are exposed to when operating in or near moving vehicle traffic. We should always operate from a defensive posture. Always consider moving vehicles as a threat to your safety. Each day, emergency personnel are exposed to motorists of varying abilities, with or without licenses, with or without legal restrictions, and driving at speeds from creeping to well beyond the speed limit. Some of these motorists are the vision impaired, the alcohol and/or drug impaired. On top of everything else, motorists will often be looking at the scene and not the road.

Nighttime operations are particularly hazardous. Visibility is reduced and the flashing of emergency lights tend to confuse motorists. Studies have shown that multiple headlights of emergency apparatus (coming from different angles at the scene) tend to blind civilian drivers as they approach.

SAFETY BENCHMARKS

Emergency personnel are at great risk while operating in or around moving traffic. There are approaches that can be taken to protect yourself and all crew members:

1. Never trust the traffic
2. Engage in proper protective parking
3. Wear orange, high visibility reflective vests
4. Reduce motorist vision impairment
5. Use traffic cones and flares

Listed below are benchmarks for safe performance when operating <u>in</u> or <u>near</u> moving vehicle traffic.

1. Always maintain an acute <u>awareness</u> of the high risk of working in or around moving traffic. Never trust moving traffic. Always look before you step! Always keep an eye on the traffic!

2. Always position apparatus to protect the scene, patients, emergency personnel, and provide a protected work area. Where possible, angle apparatus at 45 degrees away from curbside. This will direct motorist around the scene (See Figure 1). Apparatus positioning must also allow for adequate parking space for other fire apparatus (if needed), and a safe work area for emergency personnel. Allow enough distance to prevent a moving vehicle from knocking fire apparatus into the work areas.

3. At intersections, or where the incident may be near the middle of the street, two or more sides of the incident may need to be protected. Block all exposed sides. Where apparatus is in limited numbers, prioritize the blocking from the most critical to the least critical (See Figures **2**, 3 and **4**).

4. For first arriving engine companies where a charged hoseline may be needed, angle the engine so that the pump panel is "down stream," on the opposite side of on-coming traffic. This will protect the pump operator (See Figure 5).

5. The initial company officer (or Command) must <u>assess</u> the parking needs of later-arriving fire apparatus and <u>specifically direct</u> the parking and placement of these vehicles as they arrive to provide protective blocking of the scene. This officer must operate as an initial safety officer.

6. During daytime operations, leave all emergency lights on to provide warning to drivers.

7. For NIGHTTIME operations, turn OFF fire apparatus <u>headlights</u>. This will help reduce the blinding effect to approaching vehicle traffic. Other emergency lighting should be reduced to yellow lights and emergency flashers where possible.

8. Crews should exit the curbside or non-traffic side of the vehicle whenever possible.

9. Always look before stepping out of apparatus, or into any traffic areas. When walking around fire apparatus parked adjacent to moving traffic, keep an eye on traffic and walk as close to fire apparatus as possible.

10. Wear the orange safety vest any time you are operating <u>in</u> or <u>near</u> vehicle traffic.

11. When parking apparatus to protect the scene, be sure to protect the work area also. The area must be protected so that patients can be extricated, treated, moved about the scene, and loaded into Rescues safely.

12. Once enough fire apparatus have "blocked" the scene, park or stage unneeded vehicles off the street whenever possible. Bring in Rescue companies one or two at a time and park them in safe locations at the scene. This may be "down stream" from other parked apparatus, or the Rescue maybe backed at an angle into a protected loading area to prevent working in or near passing traffic. At residential medical emergencies, park Rescues in driveways for safe loading where possible. If driveways are inaccessible, park Rescues to best protect patient loading areas. (See Figures 6 and 7).

13. Place traffic cones at the scene to direct traffic. This should be initiated by the first company arriving on the scene and expanded, if needed, as later arriving companies arrive on the scene. Always place and retrieve cones while <u>facing</u> on-coming traffic.

14. Placing flares, where safe to do so, adjacent to and in combination with traffic cones for nighttime operations greatly enhances scene safety. Place flares to direct traffic where safe and appropriate to do so.

15. At major intersections a call for police response may be necessary. Provide specific direction to the police officer as to exactly what your traffic control needs are. Ensure the police are parking to protect themselves and the scene. Position Rescues to protect patient loading areas. (See Figure 8.)

FREEWAY OPERATIONS

Freeway emergencies pose a particular high risk to emergency personnel. Speeds are higher, traffic volume is significant, and civilian motorists have little opportunity to slow, stop or change lanes.

The Department of Public Safety will also have a desire to keep the freeway flowing. Where need be, the freeway can be completely shut down. This, however, rarely occurs.

For freeway emergencies, we will continue to block the scene with the first apparatus on the scene to provide a safe work area. Other companies may be used to provide additional blocking if needed.

The initial company officer, or command, must thoroughly assess the need for apparatus on the freeway and their specific positions. <u>Companies should be directed to specific parking locations</u> to protect the work area, patients, and emergency personnel.

Other apparatus should be parked downstream when possible. This provides a safe parking area.

Staging of Rescue companies off the freeway may be required. Rescues should be brought into the scene one or two at a time. A safe loading area must be established.

Traffic cones should be placed farther apart, with the last cone approximately 150 feet "upstream," to allow adequate warning to drivers. Place and retrieve cones while <u>facing</u> the traffic.

Command should establish a liaison with the Department of Public Safety as soon as possible to jointly provide a safe parking and work area and to quickly resolve the incident.

The termination of the incident must be managed with the same aggressiveness as initial actions. Crews, apparatus, and equipment must be removed from the freeway promptly, to reduce exposure to moving traffic.

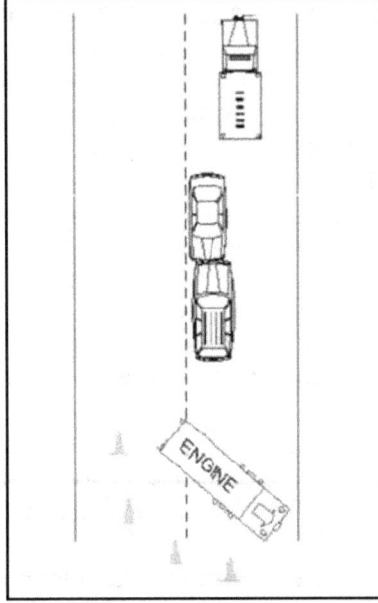

Figure 1. Where possible, angle apparatus at a 45-degree angle from the curb.

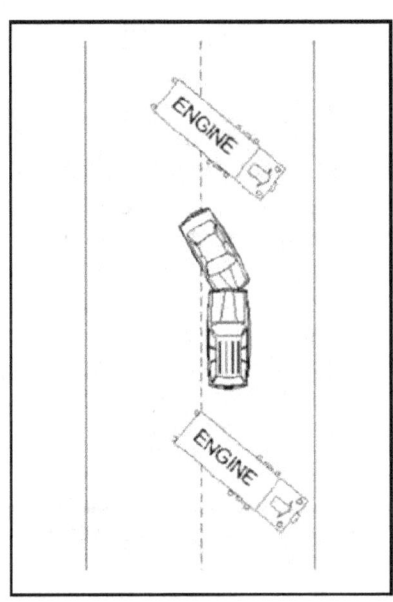

Figure 2. Prioritize placement of the apparatus by blocking from the most critical to the least critical side.

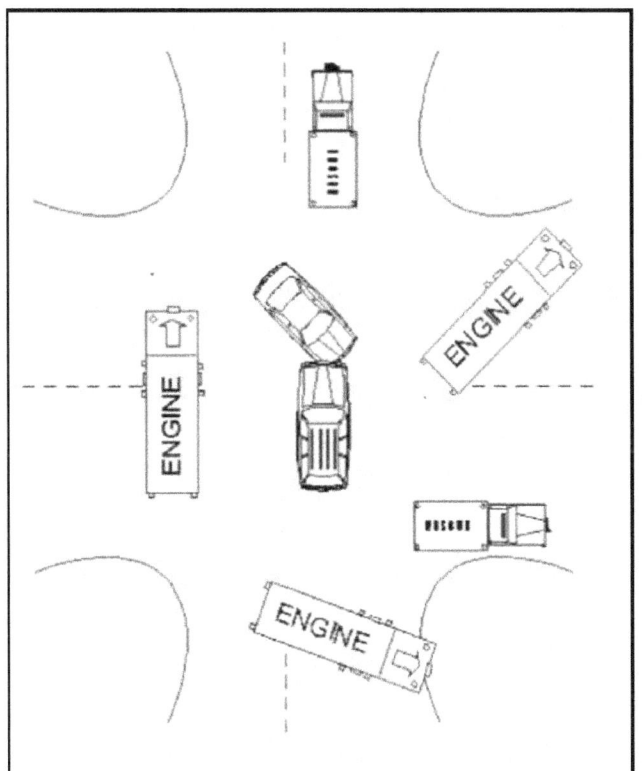

Figure 3. Often times two or more sides may need to be protected.

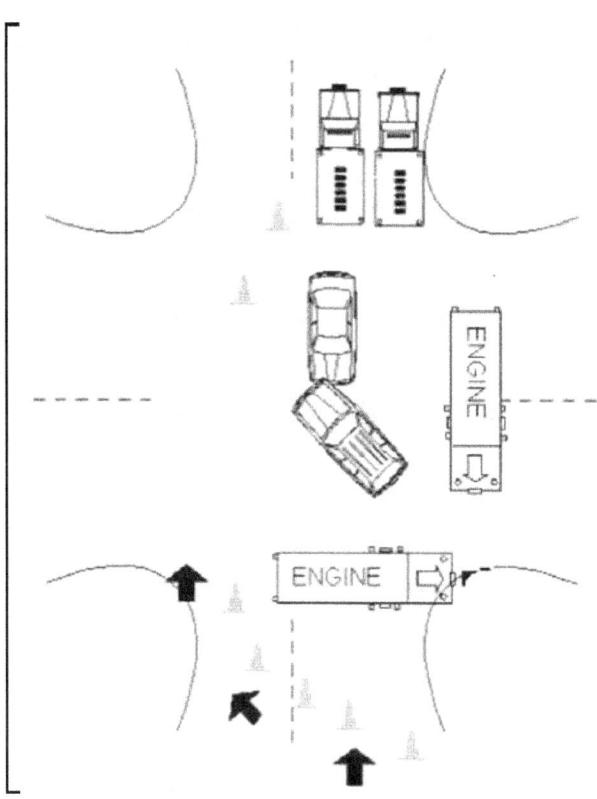

Figure 4. This collision scene must be protected from three sides.

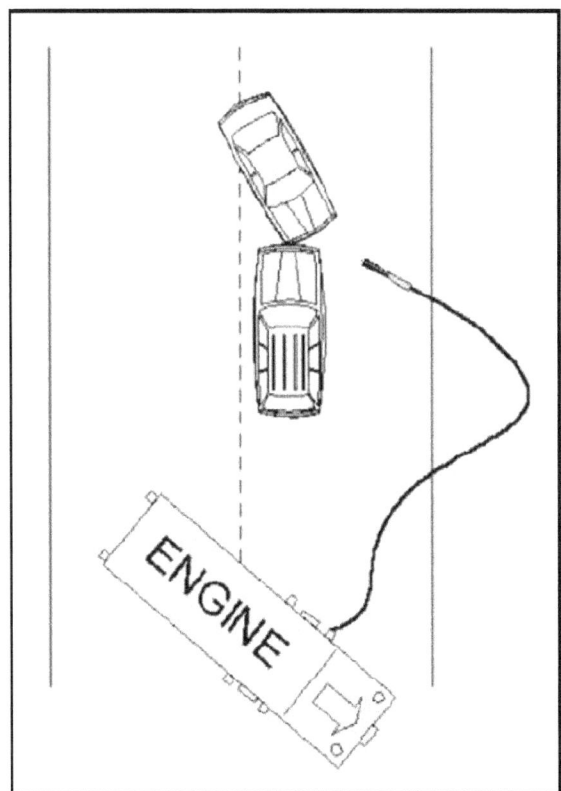

Figure 5. To protect pump operator, position apparatus with the pump panel on the opposite side of on-coming traffic.

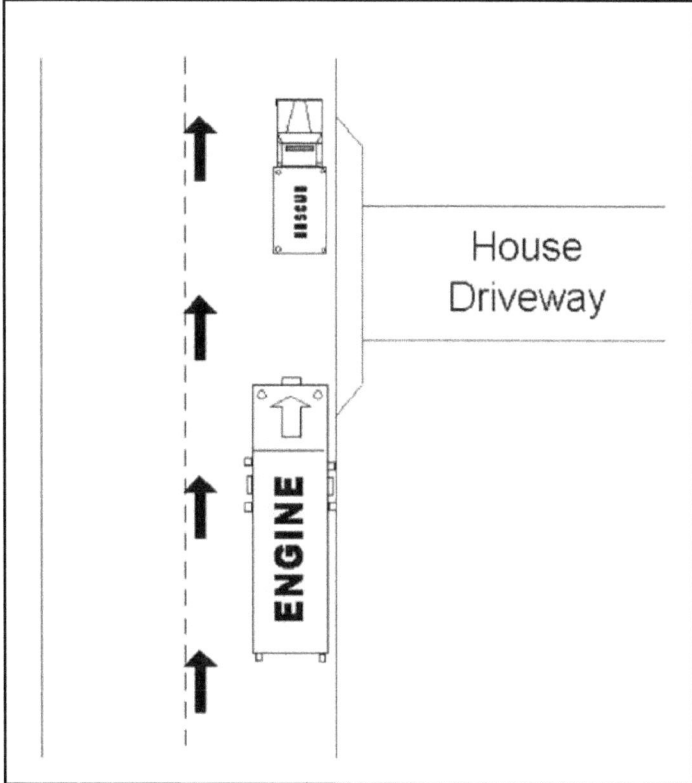

Figure 6. Where possible, park rescues in driveways or position rescue to protect patient loading area.

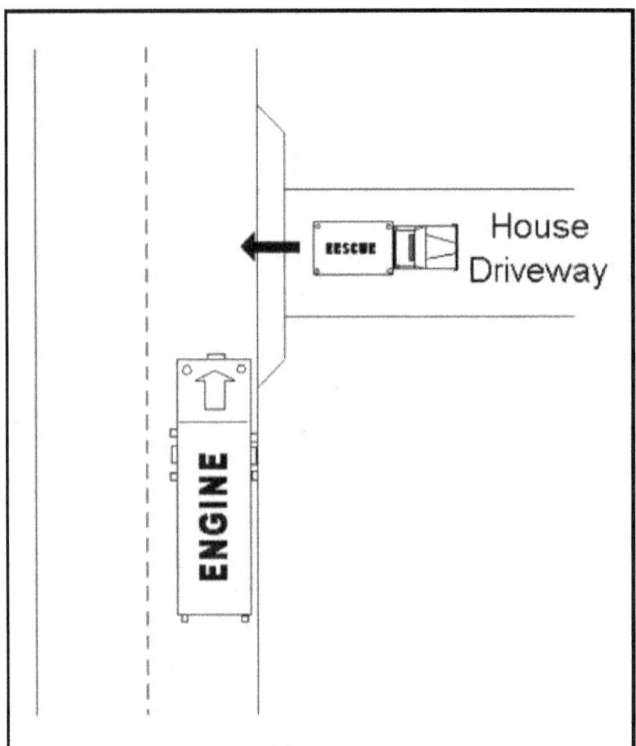

Figure 7. The patient loading area is protected when the ambulance is parked in the driveway.

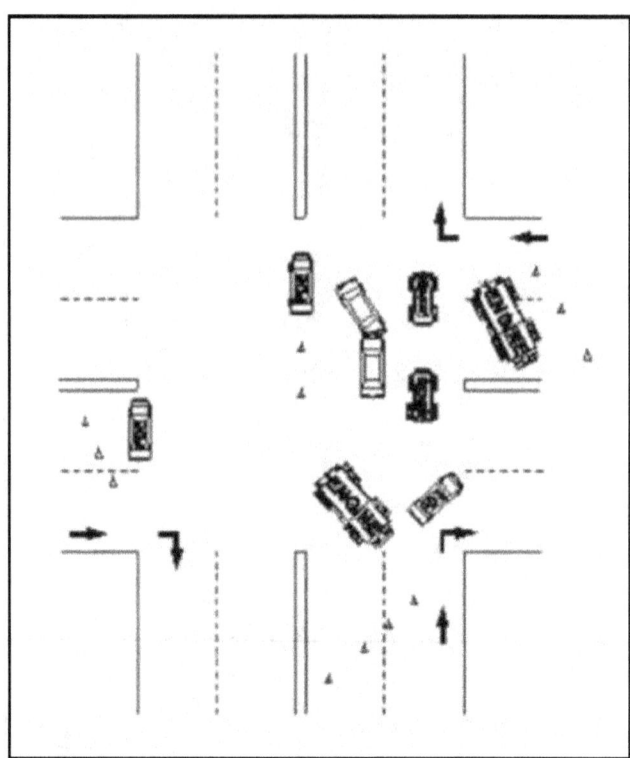

Figure 8. Provide specific direction to police as to what traffic control needs you have. Position rescues to protect patient loading areas.

PHOENIX FIRE DEPARTMENT NATURE/DISPATCH CODES

C NATURE DESCRIPTION			
ALMMNT	FIRE ALARM MAINTENANCE	2 PVT	PRIVATE AMBULANCE
ALMTST	FIRE ALARM TEST	2 REFRIG	CHECK REFRIG * AOI
BARHOS	BARRICADE HOSTAGE	2 SERV	SERVICE CALL * AOI
BBQING	BARBEQUE OPERATION	2 SLAMB	PRIVATE AMBULANCE
BLEED	CONTROL BLEEDING	2 SMOKEO	SMOKE IN AREA
CKELEC	CHECK ELECTRICAL	2 SNAKE	SNAKE REMOVAL * AOI
CKFOUT	CHECK FIRE OUT	2 SV	SERVICE VAN REQUEST
CKHAZ	CHECK HAZARD	2 TEST	THIS IS ONLY A TEST
CKWELF	CHECK WELFARE	2 UTIL	SERVICE CALL
CSPINE	CSPINE STABILIZATION		
DL	DRUG LAB	3 1000	CODE 1000
DRILL	FIRE DRILL	3 962	962
FUEL	FUEL SPILL	3 962A	962
FUELF	FUEL SPILL	3 962 BC	9C2 INV BICYCLE
FUELH	FUEL SPILL	3 962F	962 WITH FIRE
FUMIGT	FUMIGATION OPERATION	3 962HM	962 W/HAZ MATERIAL
GENRAT	GENERATOR OPERATION	3 962MC	962 INVOL MOTORCY
LGLBRN	LEGAL BURNING OPER	3 962P	962 INVOLVING PEDEST
LOCK	LOCK OUT	3 962R	962 ROLLOVER
MEDMNT	MED ALARM MAINTNEN	3 962W	CAR INTO CANAL
MEDTST	MEDICAL ALARM TEST	3 962X	962 EXTRICATION
PITBBQ	BARBEQUE OPERATION	3 A1	ALERT ONE
SPRMNT	SPRINKLER MAINTENANCE	3 A2	ALERT TWO
SPRTST	SPRINKLER TEST	3 A21A	ALERT TWO
SPRTST	SPRINKLER TEST	3 A3	ALERT THREE
2 ACCID	FIRE DEPT VEH ACCIDENT	3 ABD	ABDOMINAL PAIN
2 ARSON	FIRE INVESTIGATION	3 ACUNIT	AIR CONDITIONER
2 ASSLTM	MINOR ASSAULT * AOI	3 AED	CODE
2 ASSPD	ASSIST PD * AOI	3 ALARM	FIRE ALARM INDICATIO
2 BACKM	MINOR BACK INJ * AOI	3 ALLEY	ALLEY FIRE
2 BITEM	MINOR BITE * AOI	3 ALLRG	ALLERGIC REACTION
2 BURNM	MINOR BURN* AOI	3 ALOC	ALTERED LVL OF CONSC
2 CKBEE	CHECK BEES * AOI	3 APPLIA	APPLIANCE FIRE
2 COMCAR	CRISIS CARE	3 APT	APARTMETN FIRE
2 CONCT	CONNECTOR CALL	3 APT1A	APARTMENT FIRE
2 CROWD	CHECK CROWDING * AOI	3 ASSLT	ASSAULT
2 ILLEG	ILLEGAL BURNING	3 ASSLTS ASSLT	* STAGE FOR PD
2 INFODR	DROWNING HOTLINE	3 BACK	BACK PROBLEM
2 INJM	MINOR INJURY * AOI	3 BBQ	BBQ OUT OF CONT
2 INV	FIRE INVESTIGATION	3 BEE	BEE ASSIGNMNT
2 LAD	SERVICE CALL	3 BITE	ANIMAL BITE

2 LARC	LARC REQUEST	3 BOATA	BOAT ACCIDENT
2 MISSIN	MISSING PERSON * AOI	3 BOATF	BOAT FIRE
2 NOSEND	NON-DISPATCHED CALL	3 BR	BRUSH FIRE
2 ODOR	CHECK AN ODOR * AOI	3 BR1A	BRUSH FIRE
2 OH	OPEN HYDRANT * AOI	3 BRST	BRUSH FIRE
2 PDCCU	ASSIST PD	3 BURN	BURN INJURY
2 POOL	CHECK POOL * AOI	3 CAR	CAR FIRE
3 CARA	CAR-ABOVE GRND	3 HMED21	HAZARDOUS SITUATION
3 CAREXP	ACRE FIRE WITH EXPOSU	3 HOUS1A	HOUSE FIRE
3 CARU	CARE FIRE UNDERGRND	3 HOUSE	HOUSE FIRE
3 CB	CHILDBIRTH	3 HR1A	HIGH RISE FIRE
3 CHOKE	PERSON CHOKING	3 HR2-1	HIRISE 3-1 RESPONSE
3 CHOKEC	CHILD CHOKING	3 HR3-1	HIRISE 3-1 RESPONSE
3 CHOKEP	INFANT CHOKING	3 HYRES	HEAVY RESCUE
3 CODE	CODE	3 ILL	ILL PERSON
3 CODEC	CHILD CODE	3 INJ	INJURED PERSON
3 CODEP	INFANT COCE	3 INJX	INJURED PERSON
3 COMM	COMMERCIAL STRUCTURE	3 INTB	INTERNAL BLEEDING
3 COVAMB	COVER FD AMBULANCE	3 MA	MUTUAL AID
3 CP	CHEST PAIN	3 MAT	MATERNITY PROBLEMS
3 CRASH	AIRCRAFT DOWN	3 MED1A	MEDICAL RESPONSE
3 CSPACE	CONFINED SPACE RES	3 MED2-1	MEDICAL RESPONSE
3 CUT	CUTTING	3 MED3-1	MEDICAL RESPONSE
3 CVA	STROKE	3 MEDALM	MEDICAL ALARM
3 DB	DIFFICULTY BREATHING	3 MMRS	METRO MEDICAL RESPO
3 DEBRIS	DEBRIS FIRE	3 MOBILE	MOBILE HOME FIRE
3 DIAB	DIABETIC PROBLEM	3 MTNRES	MOUNTAIN RESCUE
3 DOWN	PERSON DOWN	3 NBC	HAZARDOUS SITUATION
3 DR	DROWNING	3 NBCNIT	HAZARDOUS SITUATION
3 DR2	DROWNING * 2 VICTIMS	3 NOSE	NOSE BLEED
3 DR3	DROWNING * 3 VICTIMS	3 OD	OVERDOSE
3 DRYER	DRYER FIRE	3 OVEN	OVEN FIRE
3 DUMP	DUMPSTER FIRE	3 POISN	POISON INGESTION
3 ELEC	ELECTROCUTION	3 POLE	POLE FIRE
3 ENG	SERVICE CALL	3 RES	RESCUE CALL
3 EYE	EYE INJURY	3 RES1A	RESCUE CALL
3 FALL	FALL INJURY	3 RES2-1	RESCUE CALL
3 FENCE	FENCE FIRE	3 RES3-1	RESCUE CALL
3 FIELD	FIELD FIRE	3 RIC	3-1 RIC RESPONSE
3 GARAGE	GARAGE FIRE	3 RIC1A	1A RIC RESPONSE
3 GASL	NATURAL GAS LEAK	3 SEIZ	SEIZURE
3 BASM	BROKEN NAT GAS MAIN	3 SHED	SHED
3 GASS	NATURAL GAS LEAK INS	3 SMOKEI	SMOKE INSIDE STRUCT
3 GRASS	GRASS FIRE	3 STAB	STABBING
3 GSW	GUNSHOT WOUND	3 STABS	STABBING-STG FOR PD
3 GSWS	GUNSHOT WOUND-STAGE	3 STOVE	STOVE FIRE

3 HA	HEART PROBLEMS	3 STR	STRUCTURE FIRE
3 HANG	HANGING	3 STR1A	STRUCTURE FIRE
3 HAZ	HAZARDOUS SITUATION	3 TASER	INJURED PERSON
3 HAZ1A	HAZARDOUS SITUATION	3 TRAIN	TRAIN FIRE
3 HAZ2-1	HAZARDOUS SITUATION	3 TRANSF	TRASNFORMER FIRE
3 HAZ3-1	HAZARDOUS SITUATION	3 TRASH	TRASH FIRE
3 HAZMED	HAZARDOUS SITUATION	3 TREE	TREE FIRE
3 HEAD	HEADACHE	3 TREERS	TREE RESCUE
3 HEAT	HEAT RELATED ILL	3 TRENCH	TRENCH RESCUE
3 HMED1A	HAZARDOUS SITUATION	3 TRK	TRUCK FIRE
3 TRKA	TRUCK ABOVE GRND	3 VEH	VEHICLE FIRE
3 TRKEXP	TRUCK FIRE WITH EXPO	3 VEHA	VEHICLE ABOVE GRND
3 TRKU	TRUCK UNDERGRND	3 VEHEXP	VEHICLE FIRE W/EXPOS
3 UNC	UNCONSCIOUS PERSON	3 VEHU	VEHICLE UNDERGRND
3 UNKF	UNKNOWN FIRE	3 WATER	WATER RESCUE
3 UNKM	UNKNOWN MEDICAL	3 WIRES	CHECK LINES DOWN

RESPONSE MATRICES

Part 1: Virginia Beach Fire and Rescue — Fire Response Matrix

Part 2: Salt Lake Valley — Fire Response Matrix Mass Casualty Response

VIRGINIA BEACH FIRE DEPARTMENT
RESPONSE RECOMMENDATION MATRIX

CALL TYPES	TYPE CODES	DIVISION CHIEF	BATTALION CHIEF	ENGINE	TRUCK	SQUAD	FIRE BOAT	WORKING INCIDENT	AMBULANCE	NOTES
Chimney Fire	HACF			1	1					Any indication of smoke or flames and calls shall be classified as the appropriate structure
Odor of Smoke	ODSM			1						Any indication of smoke or flames and calls shall be classified as the appropriate structure
Elevator Emergency	ELEV			1	1					
MARINE INCIDENTS										
Boat on Land	NBOA		1	2	1					
Boat in Water	NBOW		1	1			1	FSTAFF, PIO, Safety, PU, Duty DC, Salvage Truck, BC, Closest Squad	1 on working incidents	
Marina Fire	NMAR		1	3	1	FSQ3	1	FSTAFF, PIO, Safety, PU, Duty DC, Salvage, BC	1 on working incidents	
Swimmer in Distress/Injury on Boat	SWIM			1			1		1	Fire Boat in Company 1 and 4 area only – others as requested.
Ship Fire	NSHP		1	1				FSTAFF, PIO, Safety, PU, Duty DC, Salvage, BC	1 on working incidents	Battalion Chief shall evaluate information and determine upgraded response.
STRUCTURE FIRES										
1 or 2 Family	SRES		1 1 1	3 4 3	1 1 2	1		FSTAFF, PIO, Safety, PU, Duty DC, Salvage, BC, closest Squad if not on initial response	1 on working incidents	Response package based on closest units.
Multi-Family	SAPT		1 1 1	3 4 3	1 1 2	1		FSTAFF, PIO, Safety, PU, Duty DC, Salvage, BC, closest	1 on working incidents	Response package based on closest units.

CALL TYPES	TYPE CODES	DIVISION CHIEF	BATTALION CHIEF	ENGINE	TRUCK	SQUAD	FIRE BOAT	WORKING INCIDENT	AMBULANCE	NOTES
Commercial Fire	SCOM		1 1 1	3 4 3	1 1 2	1		FSTAFF, PIO, Safety, PU, Duty DC, Salvage, BC, closest Squad if not on initial response	1 on working incidents	Response package based on closest units.
Mall Fire	SMAL		1 1	3 4	2 2	1		FSTAFF, PIO, Safety, PU, Duty DC, Salvage, BC, closest Squad if not on initial response	1 on working incidents	Response package based on closest units.
High-Rise/4 Stories Plus	SHRS		1 1	3 4	2 2	1		FSTAFF, PIO, Safety, PU, Duty DC, Salvage, BC, closest Squad if not on initial dispatch	1 on working incident	Response package based on closest units.
Hospital/Nursing Homes	SHSP		1 1	3 4	2 2	1		FSTAFF, PIO, Safety, PU, Duty DC, Salvage, BC, closest Squad if not on initial dispatch	2 on working incident	Response package based on closest units.
AIRCRAFT INCIDENTS										
Crash – On Land	AAIR		1	3	1	1		FSTAFF, PIO, Safety, PU, Duty DC, Salvage, BC	Per EMS Matrix	
Crash – In Water	AAIR		1	1			1	FSTAFF, PIO, Safety,	Per EMS Matrix	

VIRGINIA BEACH FIRE DEPARTMENT
RESPONSE RECOMMENDATION MATRIX

CALL TYPES	TYPE CODES	DIVISION CHIEF	BATTALION CHIEF	ENGINE	TRUCK	SQUAD	FIRE BOAT	WORKING INCIDENT	AMBULANCE	NOTES
NAS Oceana Crash (large body)	NASC		1	3	2			FSTAFF, PIO, Safety, PU, Duty DC, Salvage, BC	Per EMS Matrix	Response based on direction from Oceana Command
NAS Oceana – In Flight Emergency	NASE			1						
Airport Alert – Norfolk	ASTA			1	1					
EXPLOSIVES										
Suspicious Device Standby	BOMB		1	1		FSQ3			1	PIO if requested by Command
Explosion	BOMB	1	1	3	1	1		FSTAFF, PIO, Safety, PU, Duty DC, Salvage, BC	1	Battalion 3. Based on comments regarding injuries, EMS may initiate MCI response
Hazardous Materials	MHAZ		1	3	1	FSQ3		FSTAFF, PIO, Safety, PU, Duty DC, Salvage, BC	1	Battalion 3, FSTAFF, EMS-5
FLAMMABLE LIQUID/GAS LEAK										
Commercial	CFLM		1	2	1	1				PIO and Safety as requested by Command
Residential	RFLM		1	2	1	1				PIO and Safety as requested by Command
FIRE ALARMS										
Residential	RALR			1	1					
Multi-Family/Commercial	MALR			1	1					
High-Rise/Mall	CALR		1	2	1					
Hospital/Nursing Home	HALR		1	2	1					
Carbon Monoxide Detector	ACOD			1 or 2	1 or 2	1				Response based on closest unit
TECHNICAL RESCUE										
Trench Collapse	TREN		1	1	2	2		FSTAF F, PIO Safety, PU, Duty DC, BC	1	Battalion 3 dispatched, EMS-5

VIRGINIA BEACH FIRE DEPARTMENT
RESPONSE RECOMMENDATION MATRIX

CALL TYPES	TYPE CODES	DIVISION CHIEF	BATTALION CHIEF	ENGINE	TRUCK	SQUAD	FIRE BOAT	WORKING INCIDENT	AMBULANCE	NOTES
TECHNICAL RESCUES										
Trench Collapse	TREN		1	1	2	2		FSTAFF, PIO, Safety, PU, Duty DC, BC	1	Battalion 3 dispatched, EMS-5
Confined Space	CONF		1	1	2	2		FSTAFF, PIO, Safety, PU, Duty DC, BC	1	Battalion 3 dispatched, EMS-5
Structural Collapse/Commercial	CCLS		1	1	1	IC Request		FSTAFF, PIO, Safety, PU, Duty DC, BC	1	Battalion 3 dispatched, EMS-5
Structural Collapse/Residential	RCLS			1				FSTAFF, PIO, Safety, PU, Duty DC, BC	1	Battalion 3 dispatched, EMS-5
High Angle	HIGH		1	1	2	FSQ10		FSTAFF, PIO, Safety, PU, Duty DC, BC	1	Battalion 3 dispatched
2ND ALARM										
2ND Alarm		1 / 1	1 / 1	3 / 2	1 / 1	1		Support 9	1 / 1	Based on closest units

SALT LAKE VALLEY FIRE RESPONSE MATRIX

CODE	BDALE	COUNTY	MIDVALE	MURRAY	SANDY	S JORDAN	SOSL	W JORDAN	W VALLEY
ALARM	1E	1E	1E	1E	1E	1E,1A	1E	1E	1E
ALRMCO	1E	1E	1E	1E	1E	1E	1E	1E	1E
ALRMR	1E	1E	1E,CS	1E	1E	1E,1A	1E	1E	1E
ALRMW	1E, 1T	1E, 1T	1E, 1T	1E, 1T	1E, 1T	1E, 1T, 1A	1E, 1T	1E, 1T	1E, 1T
FAPT	2E, 1T	3E,1T,1R,BC	3E,1T, 1R,BC, AAC	3E,1T,1R, BC	3E,1T,BC	3E,1T,1R,1A BC	2E,1T,1R, 1A,HM73, BC, VC	3E,1T,1R, BC	2E,1T,1A, HM73, BC
FARC	1E	1E	1E	1E	1E	1E, 1A	1E	1E	1E
FBBQ	1E	1E	1E	1E	1E	1E, 1A	1E	1E	1E
FBOMBT	BC (100)	1E, BC BOMB	1E	NOTIFY BC	1E, BC	1E	NOTIFY BC	1E	NOTIFY BC
FBOMB	1E BOMB	2E,1T,1R, BC,BOMB	1E,CS,BC BOMB	1E,BC BOMB	1E,BC	2E,1A,BC BOMB	HM73, BC BOMB	1E,1A,BC BOMO	BC WVBOMB
CHURCH	2E, 1T	3E,1T, 1R,BC	3E,1T, 1R,BC AAC	3E,1T,1R, BC	3E,1T,BC	3E,1T,1A BC	2E,1T,1R, 1A,HM73, BC, VC	3E,1T, 1R,BC	2E,1T,1A HM73,BC
FCOML	2E,1T	3E,1T, 1R,BC	3E,1T, 1R,BC AAC	3E,1T, 1R,BC	3E,1T,BC	3E,1T,1R 1ABC	2E,1T,1R, 1A,HM73, BC, VC	3E,1T, 1R,BC	2E,1T,1A HM73,BC
DYCARE	2E,1T	3E,1T, 1R,BC	3E,1T, 1R,BC AAC	3E,1T, 1R,BC	2E,1T,BC	3E,1T,1R, 1ABC	2E,1T,1R, 1A,HM73, BC, VC	3E,1T,, 1R,BC	2E,1T,1A HM73,BC
FDUMP	1E	1E	1E	1E	1E	1E,1A	1E	1E	1E
FELE	1E	1E	1E	1E	1E	1E,1A	1E	1E	1E
FEXMAT	1E BOMB	1E, BC BOMB	1E, CS BOMB	1E, BC BOMB	1E, BC	1E, 1A, BC,BOMB	1E, HMY3, BC, BOMB	1E, 1A BOMB	1E, HMY3, BC, BOMB
FEXPLO	2E BOMB	23, 1T, 1R, BC BOMB	2E,1T, 1R,BC AAC, BOMB	3E, 1T,1R, BC, BOMB	2E, 1T,BC	2E,1T,1R, 1A,BC, BOMB	2E,1T,1R,1A HM73, BC, VC, BOMB	3E,1T, 1R,BC	2E,1T,1A HM73, BC WVBOMB
FFENCE	1E	1E	1E	1E	1E	1E,1A	1E	1E	1E
FFIELD	1E	1E	1E,1X,BC,CS	1E	1E	1E,1A	1E	1E,1X	1E,1X
FMISC	1E	1E	1E	1E	1E	1E	1E	1E	1E
FFUEL	1E	1E,BC,CS	1E	1E	1E	1E, 1A	1E	1E	1E, HM73
GARAGE	2E	2E,1T, 1R,BC	3E,1T, 1R,BC AAC	2E,1T, 1R,BC	2E,1T,BC	2E,1T,1R, 1A,BC	2E,1T,1R, 1A,HM73, BC, VC	3E,1T, 1R,BC	2E,1T, 1AHM73,BC
FHAZBIO	1E,HM	1E,1R BC,BOMB	1E,HM,BC	2E,BC	1E,HM,BC	1E,1A,BC	1E,HM,BC	1E,HM,BC	BC
HAZMAT	2E,HM	1EM1RM HM,BC	1E,HM,BC AAC	2E,BC	1E,HM,BC	1E,1A,BC	1E,1T,1R,1A HM73,BC,VC E42	1E,1T 1R,BC	1E,1T,1A HM73,BC
HAZFIRE	2E,HM	3E,1T,1R HM,BC	3E, 1T,1R HM,BC,AAC	3E,1T,1R,BC	2E,1T,HM BC	2E, 1T,1R, 1A,BC	2E, 1T,1R, 1A,HM73,BC, E42	3E,1T 1R,BC	2E,1T,1A HM73,BC
FHOSP	2E,1T	3E,1T,1R BC	3E,1T,1R, BC,AAC	3E,1T, 1R,BC	3E,1T,BC	3E,1T,1R, 1A,BC	2E,1T,1R, 1A,HM73, BC,VC	3E,1T, 1R,BC	2E,1T,1A HM73,BC
FHOUSE	2E,1T	3E,1T,1R BC	3E,1T,1R, BC,AAC	2E,1T 1R,BC	2E,1T,BC	2E,1T,1R, 1A,BC	2E,1T,1R, 1A,HM73, BC,VC	3E,1T 1R,BC	2E,1T,1A HM73,BC
FHYD	1E	1E	NOTIFY STA PUB WORKS	PUBLIC WORKS	1E	1E	PUBLIC WORKS	1E	GRANGER HUNTER
FILL	1E	1E	1E	1E	1E	1E,1A	1E	1E	1E
FINV	1E	1E	1E	1E	1E	1E,1A	1E	1E	1E
FLINES	1E	1E	1E	1E	1E	1E,1A	1E	1E	1E
FLOCK			1E		1E				
NATGAS	1E	1E,HM	2E,BC,CS	2E	1E	1E,1A	2E,HM73 BC	1E,1A	2E,HM3, BC
NATDIS	1E	3E,1T, 1R,BC	3E,1T,1R BC,AAC	1E,BC	2E,1T,BC	1E,1A	HM73,BC VC,MCI, COM7	3E,1T, 1R,BC	3E,1T, HM73,BC. MCI,COM7
FPLANE	2E,1A	3E,1T, 1R,BC	3E,1T,1R, 1A,BC,AAC	2E,1T,1R 1A,BC	2E,1T,1A 1HR,BC MASS	2E,1T,1R 1A,BC	3E,1T,1R,1A HM73,BC,VC SLCCRASH	3E,1T,1R 1A,BC	3E,1T, HM73,BC, SLCCRASH
PLANEST	1E	1E	1E	1E	1E	1E,1A	1E	3E,1T,1R	1E

SALT LAKE VALLEY FIRE RESPONSE MATRIX

	BDALE	COUNTY	MIDVALE	MURRAY	SANDY	S JORDAN	SOSL	W JORDAN	W VALLEY
POLICE	1E	1E	1E	1E	1E	1E	1E	1E	1E
PROPNE	1E	1E	3E,BC,CS	2E	1E	1E,1A	2E,HM73,BC	1E,1A	2E,HM73,BC
FPUB	1E	1E	1E	1E	1E	1E	1E	1E	1E
RESCUE	2E,1A	1E,T4,RE4 1R,1A,BC	1T,1R 1A,CS	1E,1T,1R	1E,1T, HR,BC	2E,1T1R 1A,BC	2E,1T,2R 2A,HM73 1HR,BC,VC	1E,1T 1R,BC	2E,1T.2R 2A,HM73,S41 BC
FROOF	1E	3E,1T 1R,BC	1E,1T,1R BC,CS	1E,BC	1E,1T HR,BC	2E,1T1R 1A,BC	2E,1T,2R 2A,MH73,1HR BC,VC	3E,1T, 1R.BC	2E,1T.2R 2A,HM73,S41 BC
FTRASH	1E	1E	1E	1E	1E	1E,1A	1E	1E	1E
SCHOOL	2E,1T	3E,1T 1R,BC	3E,1T,1R BC,AAC	3E,1T, 1R,BC	3E,1T,BC	3E,1T,1R 1A,BC	2E,1T,1R 1A,HM73, BC,VC	3E,1T 1R,BC	2E,1T,1A HM73,BC
FSHED	2E,1T	2E,1T, 1R.BC	3E,1T,1R BC,CS	2E,BC	2E,1T,BC	2E,1T, 1A,BC	1E,1T BC,VC	3E,1T,1R	1E,1T,BC
FSMOKE	1E	1E	1E	1E	1E	1E,1A	1E	1E	1E
STBY M	1E	1E	1E	1E	1E	1E	1E	1E	1E
STBY F	1E	1E	1E	1E	1E	1E	1E	1E	1E
FSTRUC	2E,1T	3E,1T 1R.BC	3E,1T,1R BC,AAC	3E,1T 1R,BC	3E,1T,BC	3E,1T,1R 1A,BC	2E,1T,1R 1A,HM73, BC,VC	3E,1T 1R.BC	2E,1T.1A HM73,BC
SWFTW	1E	1E, S&R	1WR	4WR,1AMBC	1E	1E	1E	1E	1E
FTREE	1E	1E	1E	1E	1E	1E,1A,	1E	1E,1X	1E
FTRUCK	2E	2E	1E,1T BC,CS	2E	2E,BC	2E,1A,BC	2E,BC,VC	1E,1T,1A	2E,BC
FVEH	1E	1E	1E	1E	1E	1E,1A	1E	1E	1E
FWATER	1E	1E	1E, PUB WRK	1E	1E	1E, PUB WRK	PUB WRK	1E	PUB WRK
			CS=SMD STAFF						

BRUSH FIRE TASK FORCE

	FIELD 1	FIELD 2	FIELD 3	FILED 4	FILED 5	FIELD 6		
	2X	3X*	2WT*	3WT*	4X*	4WT*		
		*include lesser response(s) unless previously dispatched						

ADDITIONAL INFORMATION

	BDALE	COUNTY	MIDVALE	MURRAY	SANDY	S JORDAN	SOSL	W JORDAN	W VALLEY
ADD ALARMS	SAME RESPONSE AS ORIGINAL DISPATCH								
TEST/WAKEUP	1900	0600	0700	0600	NONE	NONE	0600	0700	0600

SPECIALIZED EQUIPMENT

	BDALE	COUNTY	MIDVALE	MURRAY	SANDY	S JORDAN	SOSL	W JORDAN	W VALLEY
EXTRICATION	S101	E2,T4,T5	RE21, T22	T81	T31,HR33	T or E61	S41	T52, E54	RE71,HM73

SALT LAKE VALLEY MASS CASUALTY RESPONSE MATRIX

Response

The following table illustrates response assignments for a mass casualty event:

Call Type	Assignment
First Alarm Mass Casualty 10-15 Patients	1 - IMT, 1- Extrication unit, 2 -BLS Engines, 3 - ALS Units, 1 - Ambulance Strike Team, 1 - Mass Casualty Trailer (Standby) 2 - Helicopter (Standby), Command Post
Second Alarm Mass Casualty 16-30 Patients	*First Alarm Response* & 2 – Chief Officers 1 – Extrication unit, 2 – BLS Engines, 3 – ALS Units, 1 – Ambulance Strike Team, 1 – Mass Casualty Trailer, 1 – Mass Transit Unit Total: • Chief Officers – 6· • Extrication Units – 2 • BLS Engines – 4· • ALS Units – 6 • Ambulances – 10 • Mass Transit Unit – 1
Third Alarm Mass Casualty 30 < Patients	*Second Alarm Response* & 2 Chief Officers, 1 – Extrication unit, 2 – BLS Engines, 3 – ALS Units, 1 – Ambulance Strike Team, 1 – Mass Transit Unit Total: • Chief Officers – 8· • Extrication Units – 3 • BLS Engines – 6 • ALS Units – 9 • Ambulances – 15 • Mass Transit Unit – 2

Note: The Incident Commander has the discretion to modify this response protocol, to meet the unique needs of the incident.

Also, multiple function equipment (ALS/Transport/Fire Engine) will only count as a single function unit for dispatch purposes.

AMERICAN AMBULANCE ASSOCIATION

Developed by: AAA MMTS subcommittee of the Professional Standards and Research Committee

AAA BEST PRACTICE FOR EMS DRIVING OUTLINE
Elements:
- Employee selection
- Driver training and education
- Monitoring driving performance
- Company driving policies
- Prioritizing emergency responses
- Ambulance selection
- Ambulance maintenance
- Data collection
- Other factors

BEST PRACTICE — EMPLOYEE SELECTION
Interview process focuses on:
- Attitude
- Reference checks
- Verification of driving record
- Skill level
- Training
- Safety belts use
- Personality
- Public awareness
- Knowledge
- Values

Drug Testing
- Risk related behaviors — instruments to measure for new hires
- Established minimum driver qualification criteria — e.g. number of moving violations for past 36 months, no DUI convictions, reckless cites, suspensions, age, medical condition
- DOT Physical requirements — vision, hearing, etc.
- Drivers Education certificate

BEST PRACTICE — DRIVER TRAINING & EDUCATION
- Formalized classroom program — low force driving concepts
- Certified instructor
- On the road training in training vehicle
- Testing — knowledge and skill based
- Training documentation & certification

BEST PRACTICE — MONITORING DRIVING PERFORMANCE
- On-board computer monitoring system or equivalent
- Peer monitoring by partner
- Supervisory monitoring
- Automatic monitoring of both personal and on-the-job driving for traffic violations
- Other monitoring devices

BEST PRACTICE — COMPANY DRIVING POLICIES
- Nomenclature — "crashes" & "collisions" vs."accidents"
- Crash investigation committee
- Seat belts, occupant restraints and patient restraints
- Securing cabinets, equipment, and supplies in patient compartment
- Training and retraining
- Driving privileges — suspensions and revocation
- Speed limit vs. posted speed limit for priority one responses
- Impaired driver — 24-hours shift, medications, alcohol, drugs and fatigue
- Emergency mode driving, routine driving, dynamics of large vehicles, backing, low forces driving, zero tolerance standards for seat belt use and complete stops at red lights when running emergency, passing school buses with stop signs outward, etc.
- Best practices currently employed in the industry

BEST PRACTICE - PRIORITIZING EMERGENCY RESPONSES
- Medical Priority Dispatching System reduces lights and siren responses
- Use of certified Emergency Medical Dispatchers
- Dispatch policies — company and EMS system

BEST PRACTICE — AMBULANCE SELECTION
- Use of KKK standards
- Sirens and mounting locations
- Warning lights
- Loaded weight limitations
- High center of gravity problems
- Mechanical/crash worthiness
- Interior design
- Technology systems
- Preemptive devices

BEST PRACTICE — AMBULANCE MAINTENANCE
- Preventative maintenance on vehicles
- Driver's vehicle check list
- Driver's prerogatives to sideline "un-safe" ambulance policy

BEST PRACTICE — DATA COLLECTION
- Fleet miles
- Number of vehicles
- Adverse effects
- Vehicle specification data (type of vehicle)
- Severity of crash

DRIVER TRAINING COURSES

Part 1: CDFFP Vehicle Operations Course Objectives
Part 2: Sacramento Regional Training Center Lesson Plan
Part 3: Ventura County Fire Department Driving Course Descriptions

California Department of Forestry and Fire Protection Vehicle Operations Course Objectives

BFC STUDENT HANDBOOK

2A COURSE OUTLINE

BASIC FIRE CONTROL 2A COURSE REQUIREMENTS

PHYSICAL FITNESS
PHYSICAL FITNESS TRAINING   8   hours P.T. sessions
      The student will be able to...
            Improve aerobic conditioning with cross-country hiking sessions.

PUMP OPERATIONS
PUMP THEORY AND OPERATIONS   6   hours classroom
                                             17   hours skill lab
      The student will be able to...
            Describe the functions and operation of the various pumps, gauges and controls on CDF engines.
            Demonstrate the ability to pump from tank, draft and hydrant with the various CDF engines.
            Demonstrate the procedures to follow while pumping from the tank of CDF fire engines.
            Demonstrate the procedures to follow while pumping from draft with CDF fire engines.
            Demonstrate the procedures to follow while pumping from a hydrant with CDF fire engines.

HYDRAULICS   2   hours
      The student will be able to...
            Demonstrate the ability to do simple hydraulics problems.

MOTOR VEHICLE OPERATIONS
BASIC DRIVING AND AIR BRAKES   6   hours
      The student will be able to...
            Safely and correctly drive a conventional and four wheel drive vehicle powered by a gasoline and/or diesel internal combustion engine, equipped with an Allison automatic or manual transmission.
            Properly adjust an S-Cam air brake to CDF's standards.
            Describe the procedure for checking the complete air brake system for proper operation.
            Correctly identify and complete all forms related to vehicle operation.

PREVENTATIVE MAINTENANCE   6   hours classroom
                                             7   hours field
      The student will be able to...
            Identify the components of an "A" and "B" service.
            Identify the various forms associated with an "A" and "B" service.
            Correctly perform an "A" and "B" service on a CDF engine.

EMERGENCY VEHICLE OPERATIONS   6   hours classroom
                                                   8   hours field exercise

The student will be able to...
- Identify and use the basic principles of defensive driving.
- Correctly operate a vehicle in code three mode.
- Correctly identify the limitations of and the proper use of emergency warning devices.
- Correctly identify the laws and policies that govern code three operations.

CHECK-OUT DRIVES   8   hours

The student will be able to...
- Demonstrate the ability to operate CDF engines on rural and city roadways while obeying motor vehicle laws and Department policies.

CROSS COUNTRY DRIVING   8   hours

The student will be able to...
- Demonstrate the ability to operate CDF engines in a variety of driving conditions while obeying motor vehicle laws and Departmental policies.

OFF ROAD VEHICLE OPERATIONS   2   hours classroom
                                                   9   hours field

The student will be able to...
- Provide the student with knowledge of procedures and safety considerations pertaining to off-road operations of CDF vehicles.
- Provide the student with a working knowledge of winching operations.
- Provide the student an opportunity to demonstrate the ability to safely operate a CDF vehicle off-road.
- Provide the student an opportunity to demonstrate the ability to safely use a winch and snatch block.

EMERGENCY/FIREGROUNDS OPERATIONS

INTRODUCTION TO I-ZONE   4   hours

The student will be able to...
- Recognize ways to properly deploy personnel and equipment during a wildland fire in an urban-interface area.
- Recognize how to triage structures for protection.
- Know what to do if the safety of the fire fighters is compromised.

MULTI-COMPANY DRILLS   8   hours

The student will be able to...
- Demonstrate teamwork and leadership skills while responding to a variety of emergencies as a Company Officer on an engine company.
- Demonstrate the proper techniques for handling any emergency they are dispatched to.

SACRAMENTO REGIONAL TRAINING CENTER LESSON PLAN

NOTE: This lesson plan applies to sedans and smaller vehicles.

III. VEHICLE PLACEMENT EXERCISES

A. ACCIDENT AVOIDANCE/LANE-CHANGE EXERCISE

1. Purpose: to show the student that, if braking is not an available option, an obstacle in the road can be avoided by displacing the vehicle <u>around</u> the obstacle.
2. The student will consider the following driving aspects:
   a. Proper instructed speed
   b. Good hand positioning
   c. Control of weight transfer/spring loading
   d. Oversteer recovery (if applicable)
   e. Good road positioning & High Visual Horizon
   f. 40 MPH speed — 45 MPH maximum
3. ALL EXERCISES WILL BE DEMONSTRATED TO THE STUDENT

B. MEDIUM SPEED SERPENTINE BACKING

1. Purpose: to have the driver properly position a vehicle while backing at a medium speed, taking into account Front End Swing, seating position and visual awareness to the rear.
2. The student will consider:
   a. Proper 3-point seating position:
      1. Turning to the right in the seat, directing vision through the rear window
      2. Left leg and foot to brace the body, prepared to brake
   b. Front End Swing is an important factor, to avoid front bumper/fender damage
   c. To account for F.E.S., the student should:
      1. If backing left, keep left side of the car close to obstacle
      2. If backing right, keep right side of the car close to obstacle
   d. Do not trust mirrors while backing, turn physically around to ensure your confidence in proper vehicle placement
   e. If unsure, get out and examine where you will be backing
   f. Turning movements should be smooth, using High Visual Horizon.

C. THRESHOLD BRAKING EXERCISE

1. Purpose: to provide the student with the most efficient manner of stopping a vehicle without using ABS.
2. Threshold Braking should be considered a "PANIC STOP"
3. Remember: ROLLING FRICTION = GOOD; SLIDING FRICTION = BAD
4. The brake pedal should be pushed sufficiently hard that the tires grip to their point of adhesion and NOT past it, so that the tires are still rolling
5. In other words: right to the THRESHOLD of sliding
6. If Rolling Friction is lost, REGAIN the Rolling Friction
   a. Slightly release pressure on brake pedal
   b. Do NOT take foot off brake
   c. Relax muscle of leg
7. The stop should be smooth and controlled, turning the vehicle as cones indicate.
8. Speed: 45 MPH maximum

D. T-BOX EXERCISE

1. Purpose: to provide the student with the ability to properly and safely maneuver a vehicle in and out of areas containing minimal space.
2. The student will consider:
   a. Proper 3-point seating position
   b. Proper stall positioning
      1. Pre-plan for any backing
      2. Factor in Front End Swing, Rear Wheel Cheat, Pivot Point
      3. Vision, when backing, should be over the right shoulder and through the rear window.
   c. Steering and throttle should be smooth and coordinated.

E. PARALLEL PARKING

1. Purpose: to provide the student with the ability to properly parallel park a standard-sized sedan, using correct backing principles.
2. Proper 3-point braced seating position, looking out rear window
3. Factor in Front End Swing while backing, and Slip angle when exiting
4. Throttle: smooth, controlled, slow speed
5. Put far delineator in alignment with center-line of trunk
6. Continue backing, when front fender is clear, turn into stall

F. 180× BOOTLEG TURNAROUND

1. Purpose: to provide a student the ability to turn a vehicle around in the most efficient manner possible so as to, for example, follow a suspect traveling in another direction.
2. The student will consider:
   a. Proper 3-point seating position
   b. Proper stall positioning
      1. Pre-plan for any backing
      2. Factor in Front End Swing, Rear Wheel Cheat, Pivot Point
   c. Vision should be over the right shoulder and through the rear window when backing
   d. Steering and throttle should be smooth and coordinated

G. STAR EXERCISE

1. Purpose: to provide the student the ability to properly and safely maneuver a vehicle in and out of a star-shaped area containing minimal space
2. The student will consider:
   a. Proper 3-point seating position
   b. Proper positioning around delineators
      1. Set yourself up properly for backing
      2. Account for Front End Swing, Rear Wheel Cheat, Pivot Point
      3. Vision, when backing, should be over the right shoulder and through the rear window when backing
      4. The left vehicle area should also be checked for proper vehicle positioning
   c. Steering and throttle should be smooth and coordinated

H. TURNAROUND MANEUVER

1. Purpose: to give the student the ability to maintain safe vehicle control while performing turn-around maneuvers.
2. The "Blind Spot" should always be checked
3. The Turnaround consists of a 3-Point turn, a "modified" bootleg, another bootleg
4. Pivot Points, Rear Wheel Cheat, Front End Swing and Slip Angle are all factors
5. Smooth and controlled throttle
6. Vision over right rear shoulder and through rear window
7. Consider proper roadway position
8. Confusing exercise: DIAGRAM ON BOARD

I. OFFSET FORWARD

1. Purpose: to allow the student, using good depth perception and visual awareness, the ability to properly place a vehicle through a series of restricted-clearance cones and delineators, while moving forward, and considering Rear Wheel Cheat
2. The student will consider that the average law enforcement vehicle is roughly 6' wide and factor this into their visual estimation
3. Good hand positioning will be used, the student will "feed the wheel," smoothly
4. Rear Wheel Cheat will be a factor to consider
5. Throttle should be steady and also smooth
6. All available safe roadway should be used, and vehicle should stay <u>on</u> the roadway
7. High Visual Horizon should be used

J. OFFSET BACKING

1. Purpose: to allow the student the ability to back around various obstacles and gain further awareness of spatial positioning.
2. The student will consider:
   a. Proper seating position
   b. Vision should be over the right shoulder when backing right, over the left shoulder and/or our the window when backing left
   c. Account for Front End Swing and Pivot Point
   d. When backing left, keep left side of car close to obstacle
   e. When backing right, keep right side of car close to obstacle
   f. Stay on roadway
   g. Minimize steering input:
      1. Do not hold turn too long
      2. Do not start turn too late
      3. Will make driver tend to over-correct

K. LOW SPEED SERPENTINE BACKING & FORWARD

1. Purpose: the student will learn to maneuver around obstacles moving backward and forward, using proper coordination of steering and throttle control to minimize weight transfer
2. Proper seating position will be used
3. Pivot Points, Rear Wheel Cheat and Front End Swing will be considered
4. Roadway positioning should be taken into account

VEHICLE PLACEMENT SAFETY PROCEDURES (INSTRUCTIONS TO STUDENTS)

1. Safety is paramount at all times!

2. Keep a High Visual Horizon and an awareness of other drivers on other portions of the course, such as Skid Pan. Be aware of the movement of Instructors and other vehicles.

3. Stage when and where indicated by Instructors only. Do not begin the exercise cycle unless told to do so by an Instructor, either by radio or in person.

4. Cones and delineators must be replaced by the drivers themselves when they occur, except in the Lane Change exercise, which will be completed first.

5. When moving forward or backward into a stall, come to within 12" of the end of the stall.

6. Seat belts will be worn at all times, but may be loosened or removed when backing only.

7. Use proper hand positioning: "9 & 3" or "8 & 4" or "10 & 2."

8. The air conditioners will remain off at all times — very hard on vehicles.

9. Radios in vehicles are not to be used except for safety problems/emergencies.

10. Two students per car, each driver will run the circuit twice, then switch drivers.

11. All hands, arms, etc. to be in the vehicle at all times — do not hang on to roll bars.

12. If students have a question at any time, they are encouraged to ASK.

13. Students failing to comply with safety procedures and/or Instructors' directions may be asked to leave the course and provide written documentation to his/her immediate supervisor as to why the course was not completed.

NOTE: EACH INSTRUCTOR WILL TAKE A CARLOAD OF STUDENTS THROUGH THE ENTIRE COURSE, PROVIDING DETAILED EXPLANATION OF THE EXERCISES, PRIOR TO PLACING STUDENTS INTO THE CARS.

THERE WILL BE ONE INSTRUCTOR DEMONSTRATION FOR THE ENTIRE CLASS, DEMONSTRATED TO THE GROUP:

SLIP ANGLE: An Instructor will stand next to the door of a training vehicle. With its wheels turned to the lock, the driver will, at idle, make a slow, full circle, stopping back at the Lecturing Instructor. Next, the driver will briskly accelerate the same car, wheels again turned to the lock, to show how Slip angle affects the radius of the turn.

VENTURA COUNTY FIRE DEPARTMENT DRIVING COURSE DESCRIPTIONS

COURSE: CLASS A TRACTOR/TRAILER
Description: This course is taken from the State of California Department of Motor Vehicles publication "California Commercial Driver Handbook". The course is taught on a one-on-one basis, and usually takes about 8 hours. For those that need additional instruction time, a group of about 4 personnel may work for an additional 4 hours. The goal is the mastery of the skills required to pass the DMV's "Class A" exam. The 3 areas of study consist of the "pre-trip," the "skills" portion, and the driving test.

Group: All firefighters and drivers, any support personnel required to obtain this class of license

Instructors: DMV certified in exam administration.

COURSE: CLASS B VEHICLE TRAINING
Description: The National Academy for Professional Driving has developed this course. It has both a classroom and field component and is a 2-day course. The classroom portion consists of everything from developing the need for drivers training to discussing cornering techniques. Entering highways and intersections, considering traction, accident avoidance, and addressing legal aspects of driving an emergency vehicle on the street are also topics that are covered. The field exercises begin the first day by teaching new skills such as "shuffle steering", slowing the hands and feet, and practicing these skills on a slalom course. The track changes the second day when braking, backing, and perception/reaction time are stressed.

Group: All firefighters, drivers, and captains.

Instructors: The lead instructor must be certified with the NAPD. The certification process takes 5 days and instructors must recertify every two years.

COURSE: CLASS C/STAFF VEHICLE DRIVER TRAINING
Description: This is a 4-hour course presented in either 4 one-hour sessions, or 1 4-hour session. The text is the National Safety Council's product "Defensive Driving Course-4." The information discussed covers driving conditions, unsafe driving behaviors, and aggressive driving. This course also has a final exam that is graded by the NSC, and participants will receive a certificate of completion from the NSC.

Group: All non-safety personnel that may be required to operate a non-emergency, non-commercial department vehicle.

Instructors: The instructors must attend a 2-day training session hosted by the NSC in order to present the course.

COURSE: TILLER OPERATOR CERTIFICATION
Description: This class consists of instruction, demonstration and hands-on training in a closed course setting as well as city street driving. There will be a minimum of 24 hours of training required during the 3-day course. The candidate must demonstrate competence in tiller operations, on the skills course and during city street driving.

Group: All personnel that are assigned to a station that houses a tillered truck. Others are also trained in order to keep a reasonable number of personnel available to operate the vehicle.

Instructors: Each of the instructors attended training with Santa Ana Fire Department.

COURSE: SQUAD DRIVER TRAINING

Description: This course is a compilation of the National Safety Council's "CEVO II Ambulance Operator Response Book," and the National Academy for Professional Driving. The NSC product is used for the classroom portion of the course and the NAPD skills portion is used for the field exercises. This is a 16-hour course and builds upon the foundation that has been laid during the class B vehicle training.

Group: Paramedics.

Instructors: NSC instructors and NAPD instructors team-teach this course.

COURSE: OFF-ROAD VEHICLE DRIVER TRAINING

Description: This course was designed using Kern County Fire's "Off-Road Driving Course." The National Safety Council's "Off-Road Driving" course is an additional component in the classroom portion of the course. This is a 1-day class, which consists of both classroom and field training. A representative group of the department's 4-wheel drive vehicle fleet is brought together for the class to provide students a chance to experience the differences between the various vehicles. The field exercise consists of operating one of the 4-wheel drive vehicles off-road in a steep terrain and narrow roads.

Group: Recruits, hand crew members, and personnel stationed at locations where 4-wheel drive vehicles are housed.

Instructors: The instructors for this course have been trained by Kern County fire in their "Off-Road Driving" course.

COURSE: DRIVING SIMULATOR TRAINING

Description: This course has been modeled after the law enforcement's "Peace Officers Standards and Training" that has a required component of simulated driving. The course takes 3-4 hours to deliver and has both a classroom and practical portion. Students learn to operate one of a number of simulated vehicles in any of 4 distinct virtual worlds. Initial training involves simple driving skills and progresses into emergency driving, which challenges even the most seasoned drivers. Pedestrian and vehicle conflict resolution and intersection analysis are stressed. Shuffle steering and lane positioning are also discussed. There is also the aspect of decision making when arriving on the scene of an incident as far as crew and scene safety are concerned, and an exchange of ideas related to tactical operations.

Group: All department personnel that operate emergency vehicles.

Instructors: Each of the instructors received initial training from Santa Ana Police Department.

SHARING LESSONS LEARNED

Part 1: CDFFP Green Sheet
Part 2: NIOSH Hazard ID, Traffic Hazards to Fire Fighters While Working Along Roadways

GREEN SHEET SUMMARY

RCC00054
(LMU02067)
Lassen-Modoc Unit
Northern Region

AUGUST 2, 2001
139 Incident
E-2280 Vehicle Accident, Injuries to four CDF employees

A Board of Review has not approved this Summary Report. It is intended as a safety and training tool, an aid to preventing future occurrences, and to inform interested parties. Because it is published on a short timeframe, the information contained herein is subject to revision as further investigation is conducted and additional information is developed.

INJURIES

The Engineer and three Firefighters were injured and transported by air and ground ambulances to local hospitals. The two Firefighters on the rear of E-2280 were treated for soft tissue injuries and released. The front seat Firefighter was transported to a Redding hospital with lacerations and possible C-spine injuries. The Engineer was transported to a Redding hospital, stabilized and then transferred to University of California Davis Medical Center for treatment of burn injuries. The driver of the log truck refused medical treatment at the scene.

SAFETY ISSUES FOR REVIEW

The two Firefighters on the rear of E-2280 were in full wildland Personal Protective Equipment (PPE) with helmets on secured with elastic chinstraps and goggles up. The helmets came off immediately upon initial impact.

The Engineer and front seat Firefighter were wearing wildland Nomex shirts, uniform T-shirts, trousers and boots. The Nomex worn by the Engineer may have minimized burn injuries.

All occupants of E-2280 were using seatbelts. The roll bar and seatbelts on E-2280 performed as designed. The roll bar, front bumper and radiator assembly appeared to have sustained only minor damage.

| Lookouts | Communications | Escape Routes | Safety Zones |

SEQUENCE OF EVENTS

The Engineer assigned to E-2280 began his shift August 2, 2001, the morning of the accident. The Engineer performed the daily inspection and found E-2280 in proper working condition.

E-2280, with a crew of four, left Bieber Fire Station at approximately 10:15 a.m. Their assignment was to check a previous fire that occurred July 31, 2001. The Adin incident (CALMU02042) was a vegetation fire apparently started by a power pole equipment failure, located approximately one mile east of State Route 139. E-2280 was to then continue on to Grasshopper Fire Station.

E-2280 proceeded east on State Hwy 299, then south on State Route 139 at Adin. There were no vehicles visible ahead of E-2280 on State Route 139. The front seat Firefighter, who had previously been to the Adin incident, was riding in the cab giving access directions.

While traveling southbound on State Route 139, the Engineer observed a log truck in the side view mirror following the engine at a considerable distance. The front seat Firefighter gave the Engineer advance notice of the left turn to access the fire. The Engineer activated the turn signal, prepared to turn left as directed and observed the log truck in the driver's side mirror approaching the rear of E-2280. The Engineer made certain the northbound lane was clear and continued with the left turn.

While negotiating the left turn, E-2280 was struck on the left rear corner by the loaded log truck. The truck tractor slid down the left side of the engine and part of the load on the trailer struck E-2280 in the left tail light area. The log truck came to rest on its left side on the east shoulder of the road.

The impact forced E-2280 to spin clockwise and slide sideways. Fuel from the ruptured diesel tank caught fire, burning the Engineer through the open driver's side window. E-2280 then rolled approximately 1° times and came to rest on the driver's side. The Engineer and the front seat Firefighter exited the cab through the windshield opening and helped the other Firefighters off the rear of the engine. The Engineer checked the welfare of his crew and log truck driver, called 911 from a nearby house and took charge of the scene until arrival of the first responding engine.

| Lookouts | Communications | Escape Routes | Safety Zones |

SUMMARY

At approximately 10:30 a.m. on Thursday, August 2, 2001, a traffic collision involving CDF Engine 2280 (E-2280) and a loaded log truck occurred on State Route 139, approximately 1.3 miles south of Adin, California. Both vehicles were traveling southbound when E-2280, enroute to check a previous fire, slowed to make a left turn, when it was struck on the left rear corner by the loaded log truck. The impact caused E-2280 to slide broadside, catch fire, and roll. E-2280 was destroyed by fire, impact and rollover damage. The loaded log truck also rolled onto its left side and came to rest on the east side of the roadway. One Limited Term Fire Apparatus Engineer (Engineer) and three Firefighters, from the Lassen-Modoc Unit, and the log truck driver were involved.

CONDITIONS

ROAD

The legal description of the accident location is S33; T39N; R9E; MDBM and was 69 feet north of mile post 139 LAS 65.50

State Route 139 is a north/south directional 2-lane asphalt road. The road surface was dry. Visibility was excellent to the north and south approximately 0.5 miles in each direction. The accident occurred where the road is straight and flat, the centerline is a broken stripe and the outsides of the lanes are marked with continuous white fogline. The road width is approximately 20 feet with each lane approximately 10 feet wide. There is a 2-inch drop from the road surface to the shoulder. The shoulders are unimproved and have a 25 percent downhill slope away from the roadway edge.

WEATHER

The weather was clear and dry and was not considered to be a factor in this accident. The data received from the Ash Valley Remote Automated Weather Station (RAWS), 14 miles southeast of Adin, (S30; T37N; R11E; MDBM) was as follows:

Time	Temperature	Relative Humidity
1009 hrs	76 degrees F	25%
1109 hrs	80 degrees F	19%

Lookouts   Communications   Escape Routes   Safety Zones

EMERGENCY VEHICLE SAFETY INITIATIVE

August 2, 2001

139 Incident – Engine 2280

NIOSH HAZARD NOTIFICATION SHEET
NIOSH TRAFFIC HAZARDS TO FIRE FIGHTERS
WHILE WORKING ALONG ROADWAYS

DESCRIPTION OF HAZARD

The number of fire fighters struck and killed by motor vehicles has dramatically increased within recent years. During the 5-year period between 1995 and 1999, 17 fire fighters were struck and killed by motorists. This represents and 89 percent increase in the number of line-of-duty deaths over the previous 5-year period (between 1990 and 1994), when 9 fire fighters were struck and killed by motor vehicles [NFPA 2000]. Under the Fire Fighter Fatality Investigation and Prevention Program, NIOSH investigated two separate incidents involving fire fighters who were struck and killed while providing emergency services along roadways during 1999 [NIOSH 1999, 2000]. These incidents and data demonstrate that hazards to the fire service are not limited to structural or wildland fires. Motorists accustomed to a clear, unobstructed roadway may not recognize and avoid closed lanes or emergency workers on or near the roadway. In some cases, conditions can reduce a motorist's ability to see and avoid fire fighters and apparatus. Some examples include weather, time of day, scene lighting (i.e., area lighting and optical warning devices), traffic speed and volume, and road configuration (i.e., hills, curves, and other obstructions that limit visibility). These hazards are not limited to the fire service alone. Other emergency service providers such as law enforcement officers, paramedics, and vehicle recovery personnel are also exposed to these hazards.

CASE STUDIES

On August 5, 1999, one fire fighter died, and a second fire fighter and another person were severely injured when they were struck by a motor vehicle that lost control on a wet and busy interstate highway [NIOSH 1999]. A heavy-rescue squad and a ladder truck had been dispatched to a single motor vehicle crash on an interstate highway. Approximately 2 minutes after they arrived on the scene and took a position to the rear of the rescue squad (protecting the initial vehicle crash scene), another car collided with the back of the ladder truck (Figure 1). While attending to the injuries of the driver who struck the ladder truck, two fire fighters and the injured driver were struck by a third car, causing one fire fighter to be fatally injured and the second fire fighter and the driver (who had collided with the back of the ladder truck) to be severely injured.On September 27, 1999, a fire fighter died after being struck by a tractor trailer truck while directing traffic along a four-lane highway [NIOSH 2000]. The victim was standing in front of an apparatus that was parked (facing north) in the outer emergency lane for the southbound traffic. The emergency lights of the apparatus were on and functioning properly at the time of the incident. He was called out to provide assistance for a neighboring fire department that had responded to a tractor-trailer crash. The initial tractor-trailer crash and subsequent fire fighter fatality occurred during a heavy rainstorm along a 1-mile stretch of a four-lane highway. Thirty-nine collisions have occurred on this 1-mile stretch of road since 1994.

RECOMMENDATIONS FOR PREVENTION

Any fire fighter working along any type of roadway runs the risk of being struck by a motorist. To prevent such incidents, NIOSH recommends that fire departments and fire fighters take the following actions:

Fire departments:

- Develop, implement, and enforce standard operating procedures (SOPs) regarding emergency operations for roadway incidents.
- Implement an incident management system to manage all emergency incidents.
- Establish a unified command for incidents that occur where multiple agencies have jurisdiction.
- Ensure that a separate incident safety officer (independent of the incident commander) is appointed.
- Develop pre-incident plans for areas that have a high rate of motor vehicle crashes.
- Establish pre-incident agreements with law enforcement and other agencies such as the highway department.
- Ensure that fire fighters are trained in safe procedures for operating in or near moving traffic
- Ensure that fire fighters wear suitable high-visibility apparel such as a strong yellow-green or orange reflecting flagger vest when operating at an emergency scene.

Firefighters

- Ensure that the fire apparatus is positioned to take advantage of topography and weather conditions (uphill and upwind) and to protect fire fighters from traffic.
- Park or stage unneeded vehicles off the roadway whenever possible.
- If police have not yet arrived at a scene involving a highway incident or fire, first control the oncoming vehicles before safely turning your attention to the emergency.
- Position yourself and any victim(s) in a secure area that maximizes your visibility to motorists when it is impossible to protect the incident scene from immediate danger.
- Use a traffic control device that maximizes your visibility to motorists when controlling traffic.

ACKNOWLEDGMENTS

The principal contributors to this publication were Mark F. McFall and Eric R. Schmidt, NIOSH. Internal review was provided by Stephanie Pratt, DSR, NIOSH. External reviews were provided by Rita Fahy, National Fire Protection Association; Barbara Hauser, County Department of Transportation, Phoenix, Arizona; Gary Morris, Phoenix Fire Department, Phoenix, Arizona; Ron Moore, Plano Fire Department, Plano, Texas; Guenther Lerch, Professional Engineer.

REFERENCES

NFPA [2000]. U.S. fire fighters struck by vehicles, 1977-1999. Quincy, MA: National Fire Protection Association. Unpublished.

NIOSH [1999]. One fire fighter died and a second fire fighter was severely injured after being struck by a motor vehicle on an interstate highway-OK. Cincinnati, OH: U.S. Department of Health and Human Services, Public Health Service, Centers for Disease Control and Prevention, National Institute for Occupational Safety and Health, DHHS (NIOSH) Publication No. 99F-27.

NIOSH [2000]. Volunteer fire fighter died after being struck by an eighteen-wheel tractor trailer truck-SC. Cincinnati, OH: U.S. Department of Health and Human Services, Public Health Service, Centers for Disease Control and Prevention, National Institute for Occupational Safety and Health, DHHS (NIOSH) Publication No. 99F-38.

BIBLIOGRAPHY

Dunn V [1992]. Safety and survival on the fireground. Saddle Brook, NJ: Fire Engineering Books and Videos.

Hall R, Adams B, eds. [1998]. Essentials of fire fighting. 4th ed. Stillwater, OK: Board of Regents, Oklahoma State University.

Kipp JD, Loflin ME [1996]. Emergency incident risk management: a safety and health perspective. New York: Van Nostrand Reinhold.

NFPA [1995]. NFPA standard on fire department incident management system. Quincy, MA: National Fire Protection Association, NFPA 1561. NFPA [1997]. NFPA standard on fire department occupational safety and health program. Quincy, MA: National Fire Protection Association, NFPA 1500.

NFPA [1997]. NFPA standard for a fire department service vehicle operations training program. Quincy, MA: National Fire Protection Association, NFPA 1451.

NFPA [1998]. NFPA standard for road tunnels, bridges, and other limited access highways. Quincy, MA: National Fire Protection Association, NFPA 502.

U.S. Department of Transportation, Federal Highway Administration [1998]. Standards and guides for traffic controls for street and highway construction, maintenance, utility, and incident management operations. Part VI of the manual on uniform traffic control devices

(MUTCD). 3rd rev. Fredericksburg, VA: American Traffic Safety Services Association.

DHHS (NIOSH) Publication No. 2001-143

For more information

NIOSH research on traffic hazards to fire fighters has recently been published in the following: McFall M [2001]. Roadside assistance. Fire Chief *45*(3):62-64.

To obtain more information about this hazard or other work place hazards

   -call NIOSH at 1-800-35-NIOSH, or

   -visit the NIOSH Web site at http://www.cdc.gov/niosh

The purpose of the NIOSH Fire Fighter Fatality Investigation and Prevention Program is to determine factors that cause or contribute to fire fighter deaths suffered in the line-of-duty, and to develop strategies for preventing similar incidents in the future. More information can be found at http://www.cdc.gov/niosh/firehome.html.

Department of Health and Human Services
Centers for Disease Control and Prevention
National Institute for Occupational Safety and Health
4676 Columbia Parkway
Cincinnati, OH 45226-1998

DRIVING COURSE SPECIFICATIONS

This section can be used to determine the driving course requirements as set forth in NFPA 1002, Standard for Fire Apparatus Driver/Operator Professional Qualifications

VENTURA COUNTY FIRE PROTECTION DISTRICT
ENGINEER CANDIDATE TASKBOOK

Training Section

FIRE APPARATUS ROAD COURSE

Operate a fire department vehicle, given a vehicle and a predetermined route on a public way that incorporates the maneuvers and features specified in the following list that the driver/operator is expected to encounter during normal operations so that the vehicle is safely operated in compliance with all applicable state and local laws, department rules and regulations.

- Four left and 4 right turns

- A straight section of urban business street or a two-lane rural road at least 1 mile in length

- One through-intersection and two intersections where a stop has to be made

- One railroad crossing

- One curve, either left or right

- A section of limited-access highway that includes a conventional ramp entrance and exit and a section of road long enough to allow two lane changes

- A downgrade steep enough and long enough to require down-shifting and braking

- An upgrade steep enough and long enough to require gear changing to maintain speed

- One underpass or a low clearance or bridge

DRIVER/OPERATOR   *Sec. 3-11: FIRE APPARATUS ROAD COURSE*
12/6/2002

Training Section

WILDLAND APPARATUS OFF ROAD COURSE

Operate a Wildland apparatus, given a predetermined route off of a public way that incorporates the maneuvers and features specified in the following list that the driver/operator is expected to encounter during normal operations, so that the vehicle is safely operated in compliance with all applicable state and local laws, department rules and regulations.

- Loose or wet soil

- Steep grades (30 percent fore and aft)

- Limited sight distance

- Blind curve

- Vehicle clearance obstacles (height, width, undercarriage, angle of approach, angle of departure)

- Limited space for turnaround

- Side slopes (20 percent side to side)

Training Section

AIRCRAFT CRASH VEHICLE OFF ROAD COURSE

Operate an Aircraft Crash Vehicle, given a predetermined route off of an improved surface that incorporates the maneuvers and features specified in the following list that the driver/operator is expected to encounter during normal operations, so that the vehicle is safely operated in compliance with all applicable state and local laws, department rules and regulations.

- Loose or wet soil

- Steep grades (30 percent fore and aft)

- Limited sight distance

- Vehicle clearance obstacles (height, width, undercarriage)

- Limited space for turnaround

- Side slopes (20 percent side to side)

MANUFACTURER CONTACT INFORMATION

NOTE: Manufacturers listed in this appendix are those specifically identified by the departments visited for this study, and the list is not meant to be all-inclusive. Listing in this appendix does not constitute endorsement by the Department of Homeland Security, Federal Emergency Management Agency, or the United States Fire Administration.

CAMERAS AND ELECTRONIC MONITORING:

Safety Vision
6650 Roxburgh Drive,
Suite 100
Houston, TX 77041

(713) 896-6600
(800) 880-8855
Fax: (713) 896-6640
http://www.safetyvision.com

Recording DriveCam Video Systems,
3954 Murphy Canyon Road #D10
San Diego, CA 92123

(866) 419-5861
Fax: (858) 430-4001
http://www.drivecam.com

On-Board Computer
Road Safety International, Inc.
3251 Grande Vista Drive
Thousand Oaks, CA 91320-119

(805) 498-9444
Fax:(805) 498-0944
http://www.roadsafety.com

OPTICAL PREEMPTION

3 M Traffic Safety and Management
3M Center, Building 0225-05-S-08
St. Paul, MN 55144-1000

(651) 733-3590
(800) 553-1380
http://www.3m.com/us/safety/tcm/products/
prodmain.jhtml
(Intelligent Transportation System
  Priority Control Systems)

Tomar Electronics
2100 W. Obispo Avenue
Gilbert, AZ 85233

(408) 497-4400
(800) 3383133
http://www.tomar.com/products/traffic
control.htm

Simulator Driving
Doron Precision Systems, Inc.
Box 400, 174 Court Street
Binghamton, NY 13902-0400

(607) 772-0478
Fax: (607) 772-6760
http://www.doronprecision.com

FAAC, Inc.
1229 Oak Valley Drive
Ann Arbor, MI 48108

(734) 761-5836
Fax:(734) 761-5368
http://www.faac.com

Tape Reflective
3 M Traffic Safety and Management
3 M Center, Building 0225-05-S-08
St. Paul, MN 55144-1000

(651) 733-3590
(800) 553-1380
http://www.3m.com/us/safety/tcm/products/
prodmain.jhtml
(Vehicle & Conspicuity Markings)

TRAINING MATERIAL

International Fire   (800) 654-4055   http://www.ifsta.org
Service Training Association
Fire Protection Publications
930 North Willis Street
Stillwater, OK 74078

National Fire Protection Association   (800) 344-3555   http://www.nfpa.org
P.O. Box 9101
Quincy, MA 02269

National Safety Council   (630) 285-1121   http://www.nsc.org
1121 Spring Lake Drive
Itasca, IL 60143-3201

National Academy for   (972) 225-7366   http://www.napd.com
Professional Driving
P.O.Box 649
Hutchins,TX 75141-0649

VFIS   (800) 233-1957   http://www.vfis.com
183 Leader Heights Road   Fax:(717) 741-3130
York, PA 17402

ACRONYM LIST

ALS	Advanced Life Support
AMOS	Advanced Mobile Operations Simulator
ANSI	American National Standards Institute
ASE	Automotive Service Excellence
AVL	Automatic Vehicle Locator
BFEO	Basic Fire Engine Operator
BLS	Basic Life Support
CAAS	Commission on Accreditation of Ambulance Services
CAD	Computer Aided Dispatch
CDFFP	California Department of Forestry and Fire Protection
CDL	Commercial Driver License
CFAI	Commission on Fire Accreditation International
CHP	California Highway Patrol
DMV	Department of Motor Vehicles
DOT	Department of Transportation
EFO	Executive Fire Officer
EMD	Emergency Medical Dispatch
EMS	Emergency Medical Services
EMT-B	Emergency Medical Technician-Basic
EVOC	Emergency Vehicle Operations Course
EVSI	Emergency Vehicle Safety Initiative
EVT	Emergency Vehicle Technician
FARS	Fatality Analysis Reporting System
FEMA	Federal Emergency Management Agency
FIRT	Freeway Incident Response Team
GPS	Global Positioning Satellite
HFEO	Heavy Fire Equipment Operator
IFSTA	International Fire Service Training Association
IMT	Incident Management Team
ISO	Insurance Services Office
ITS	Intelligent Transportation System
JPA	Joint Powers Agreement
MAIT	Multi-Disciplined Accident Investigation Team
MDT	Mobile Data Terminal
MUTCD	Manual of Uniform Traffic Control Devices
NAEMSP	National Association of EMS Physicians
NAEMSD	National Association of State EMS Directors
NFA	National Fire Academy
NFIRS	National Fire Incident Reporting System
NFPA	National Fire Protection Association
NHTSA	National Highway Traffic Safety Administration
ORIMS	Office of Risk Insurance Management
PFN	Phoenix Fire Network
POST	Peace Officer Standard Training
POV	Privately Owned Vehicle

SAS ....Simulator Adaptation Syndrome
SOP ....Standard Operating Procedure
TOC ....Traffic Operations Center
TXDOT ....Texas Department of Transportation
UDOT ....Utah Department of Transportation
US&R ....Urban Search and Rescue
USFA ....United States Fire Administration
VDOT ....Virginia Department of Transportation
VECC ....(Salt Lake) Valley Emergency Communications Center
VMS ....Variable Message Signs

www.ingramcontent.com/pod-product-compliance
Lightning Source LLC
Chambersburg PA
CBHW081131170526
45165CB00008B/2632